Unsustainable

About the author

PATRICK HOSSAY lives in New Jersey and teaches international development and environmental politics at the Richard Stockton College of New Jersey. He received his PhD from the New School for Social Research. He also directs community conservation and eco-development projects in Central America for the international volunteer group Peacework and remains active in the struggle for social and environmental justice.

Updates, further information, and sources for action may be found on the author's website, www.sustainablejustice.org

Unsustainable

A primer for global environmental and social justice

PATRICK HOSSAY

ZED BOOKS
London & New York

Unsustainable was published in 2006
by Zed Books Ltd, 7 Cynthia Street, London N1 9JF, UK,
and Room 400, 175 Fifth Avenue, New York, NY 10010, USA

www.zedbooks.co.uk

Designed and typeset in Monotype Garamond
by illuminati, Grosmont, www.illuminatibooks.co.uk
Cover designed by Andrew Corbett
Printed and bound in Malta by Gutenberg Press

Distributed in the USA exclusively by Palgrave Macmillan, a division of
St Martin's Press, LLC, 175 Fifth Avenue, New York, NY 10010

A catalogue record for this book is available from the British Library
Library of Congress Cataloging-in-Publication Data available

ISBN 1 84277 656 8 HB
ISBN 1 84277 657 6 Pb
ISBN 978 1 84277 656 8 Hb
ISBN 978 1 84277 657 5 Pb

Contents

Figures

The trouble we're in/ 1

We're in trouble. Put simply, we are destroying the natural systems on which our lives depend. The pollutants that we've pumped into the air, water, and soil have fundamentally changed the earth's ecological balance. Much of the damage is irreversible. The destruction of the earth's ozone layer, the acidification of our rain, the poisoning of our rivers, lakes and oceans, the depletion of our soil, the devastation of our forests, and large-scale extinctions intensify one another, creating a multi-pronged and devastating attack on the earth's capacity to support human life. In the short term, our treatment of the earth as a toxic waste dump will lead to mass environmental destruction and tremendous human suffering. In the long term, if unchecked, it will kill us all. We won't be the first to go; many species will proceed us into oblivion. But, make no mistake: if we do nothing, we will go.

It's tempting to oversimplify the problem, to say simply there are too many people, too much consumption, or too many factories. These easy claims don't really get at the trouble we're in. It's true, the rate of population growth is troubling; but we are capable of producing enough food for every person on the planet to be adequately fed in a way that does not destroy the global ecosystem. In fact, 78 per cent of all mal-nourished children live in countries with food surpluses![1] It is also true that overconsumption of the earth's resources is clearly a problem; but the large majority of the earth's people do not consume great amounts

1.1 / The global gap

	Life expectancy (years)	Under-5 mortality (%)	Malnutrition (%)	% living off less than $1 per day	% living off less than $2 per day
Sub-Saharan Africa	46	17	33	47	75
East Asia and Pacific	69	4	12	34	51
Latin America and Caribbean	70	3	12	11	27
World's poorest countries	50	16	38	27	57
World's wealthy countries	78	<1	–	–	–
World	67	8		21	43

Source: *Human Development Report*, United Nations Development Program, various years, available at www.undp.org; World Bank Statistical Information Management and Analysis (SIMA) database, available at www.worldbank.org.

of resources. Nearly half of the earth's population survive on less than $2 per day.[2] These people certainly are not overconsuming! Similarly, there are clearly too many industries pumping out too much pollution; but much of the world lives without industry, without adequate employment, without a livelihood, and lacking the resources necessary to survive.

To really understand the trouble we're in, we need to ask not only 'what's being destroyed?' but also 'by whom?,' 'how?' and 'why?' A closer look will reveal that the current system is physically unsustainable because we are consuming far more resources than the earth can sustainably provide, and we are producing massive waste and pollution that the earth's ecosystem simply cannot continue to absorb. The current system is also morally unsustainable because this environmental devastation is being wrought by a small, privileged minority of the earth's residents who live in excess, consuming huge amounts of the earth's resources, while a large number of the planet's population go without the basic necessities of life. These aren't separate problems: the destruction of the global ecosystem and the violent inequity in the distribution of wealth and resources are two sides of the same coin; we cannot address one without addressing the other.

The moral injustice of the current world order is uncontestable. Roughly one in four people in the world lack adequate food and other necessities of life.[3] The average person in the world's most impoverished countries

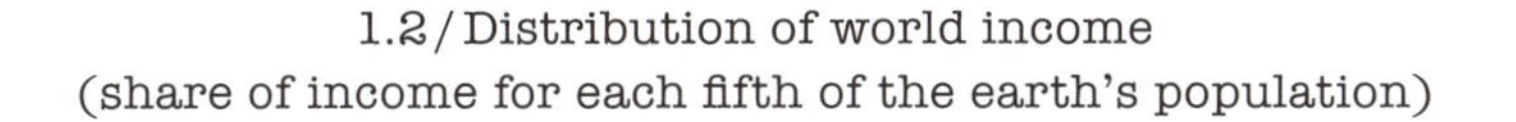
1.2 / Distribution of world income
(share of income for each fifth of the earth's population)

Source: *Human Development Report 1990*, United Nations Development Program, 1990, available at www.undp.org.

will die 27 years younger than the average person in the world's richest countries.[4] In fact, less than half of those born in the world's poorest countries can expect to live to the age of 65.[5] In southern Africa, an infant has less than a 1 in 7 chance of reaching its fifth birthday.[6] In the most impoverished areas of the world, over 40 per cent of infants are underweight; and 10 per cent of infants will die in their first year of life.[7] While these conditions are improving in some countries, in about an equal number of countries these conditions are worsening.

The gap between the world's rich and the poor is dramatic. The 20 per cent of the earth's people who live in wealthiest countries consume 58 per cent of the world's energy; the poorest 20 per cent consume 5 per cent. The wealthiest 20 per cent consume 84 per cent of all paper products; the poorest consume 1 per cent. The wealthiest 20 per cent own 87 per cent of the world's vehicles; the poorest 20 per cent have less than 1 per cent.[8] More disturbing still, this gap is getting larger. The gap in access to energy and natural resources has more than doubled in the past three decades.[9] In 1970, the number of infants who died before their first birthday was around five times higher in the world's poor countries than it was in the wealthy countries of the world; it is now about ten times higher.[10]

Not only is this expanding gap between the world's poor and rich morally intolerable, as I argue throughout this book and this chapter in particular; this simply cannot be the basis for an environmentally sustainable world. The consumption by the world's rich minority and the associated gross disregard for the earth's ecological balance have brought us to a point where business as usual simply cannot continue. Even if we in the wealthy countries of the world choose to ignore the tremendous suffering

1.3 / Unequal global shares in private consumption

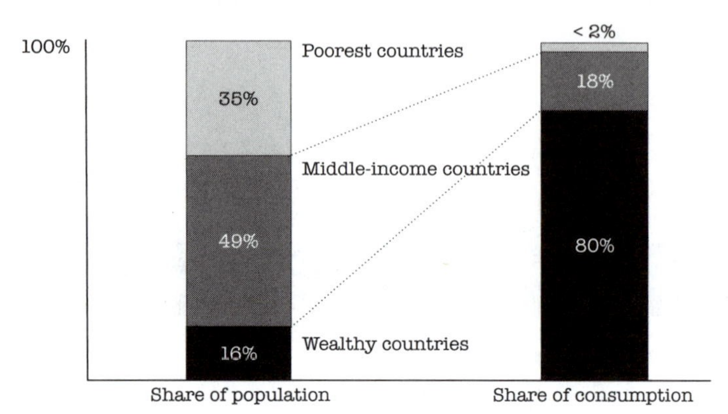

Source: UNDP, *Human Development Report, 2001*, p. 17.

of the earth's majority, even if we agree to overlook the fundamental injustice of the current order, we cannot choose to disregard the physical consequences of this system: the current world order is destroying the earth's capacity to support human life. Closing our eyes to this fact will not make it go away.

Hotter for some than for others

The environmental issue that has received the most attention recently is probably our destruction of the earth's atmosphere. By pumping millions of tonnes of pollutants into the atmosphere every year for decades, we have irrevocably altered the air around us. Even if we stopped now, the effects will be tragic. But we are not stopping; instead we are dramatically increasing our destruction of the atmosphere. Even the most cautious scientists agree that the result will be an irreversible, sweeping, and destructive change in the earth's climate.

Despite initial skepticism and organized efforts by the fossil fuel industry to deny it, the reality of global warming is now absolutely clear. In 1988, the director of NASA's Godderd Institute announced to a Senate committee that global warming had clearly begun. Corporate leaders and the scientists that they hired objected. Two years later, a UN panel composed of two thousand of the world's leading climate scientists made the same declaration: the world's temperature was clearly rising, and human-made pollution was the cause. Despite the denials of oil company executives and the contrived research they funded, the evidence was clear and mounting. In 2001, when the Bush administration asked a panel of prominent and diverse scientists for help in sorting out the science of climate change, even the most skeptical on the panel agreed that 'serious adverse societal and ecological impacts by the end of this century' were likely.[11] The Bush administration's own Environmental Protection Agency (EPA) – certainly not environmental radicals – issued a clear warning. While understated, an EPA report in 2002 was forced to admit not only that the atmosphere was being destroyed by the burning of fossil fuels but also that the resulting 'disappearance or fragmentation of natural ecosystems are likely to be costly or impossible to replace.'[12]

Global warming can be likened to sitting in a car in the sun. The sun's heat enters through the windshield but is unable to escape the same way because it has changed to longer heat wavelengths that cannot penetrate the glass. So, inside the car it gets hotter and hotter. The Earth's atmosphere creates the same effect; and as we continue to increase the amount of polluting carbon dioxide in the atmosphere, it has the same result as rolling up the windows in a hot car – less and less heat is able to escape. Eventually, as the car gets increasingly hot, the situation gets more serious. Unlike the inside of the car, the effects on our climate won't be uniform. Some areas will get considerably hotter for sure, but some will become dryer; others will become wetter. Nevertheless, while the effect is more complicated in the case of the earth's atmosphere than in the car, the outcome is the same: if we do not reduce our production of pollutants, things will get serious, fast.

Several pollutants contribute to global warming, some directly and some indirectly. The largest share of out total 'greenhouse gas' production is composed of carbon dioxide. Of course, carbon dioxide is a natural component of the earth's atmosphere. However, principally through our combustion of fossil fuels in cars, aircraft, power plants, and factories, we have raised the amount of carbon dioxide in the environment by well over one-third since the start of the Industrial Revolution. As the graph indicates, the rate of increase is now staggering.

Methane, a product of the breakdown of biological material, is also a major contributor to global warming. The largest sources are the decomposition of waste in landfills and the gas produced by livestock – cow farts. While this may not initially appear to be a major concern, livestock produce nearly 10 million tonnes of methane each year. Landfills produce even more. And methane is nearly twenty-five times more effective at trapping heat in the atmosphere than carbon dioxide. The level of methane in the

1.4 / Carbon dioxide emissions (billion tonnes)

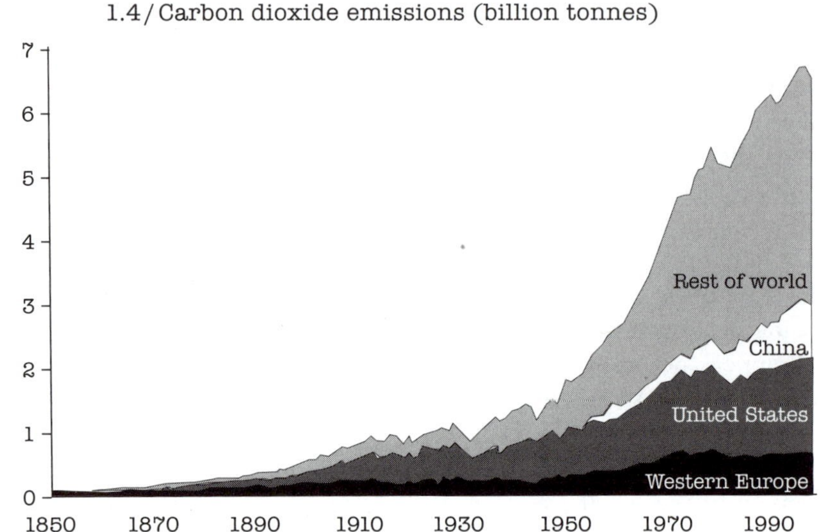

Source: Carbon Dioxide Information Analysis Center, US Department of Energy, data available at http://cdiac.esd.ornl.gov.

air is now 145 per cent over natural levels. In fact, since not all gases have an equal effect on the environment, a little bit of the wrong gas can have big consequences. Certain hydrofluorocarbons (HFC-23) are up to twelve thousand times more effective than carbon dioxide at trapping heat.[13] And sulfate aerosol (SF_6) has double the heat-trapping capacity of even the most damaging hydrofluorocarbons. Still, the largest contributor to global warming is clearly carbon dioxide. The United States alone pumps over 10 billion tonnes of carbon dioxide into the atmosphere every year.[14] And, as our consumption of energy and our use of gas-guzzling automobiles increase, so does our destruction of the air around us.

Some have wondered if the recent rise in average global temperature is a result of natural fluctuations. Mounting evidence indicates that this is very unlikely. By measuring the width of tree rings, we can calculate temperature changes around the world going back thousands of years. The composition of snow layers in glaciers, pollen abundance in lake sediments, and other gauges, allow us to confirm these measures. Recent research has allowed us to eliminate other possible causes of temperature rise such as volcanic eruptions or changes in solar radiation. In the end, the overwhelming weight of the evidence indicates that the current rise in temperature is not the result of normal historical fluctuations.[15]

In short, global warming is here. If we're lucky, over the next century alone, we can expect the average temperatures to continue to rise by 1.5 to 6 degrees Centigrade. As that happens, a marked increase in severe

1.5 / Global temperature change (°C)

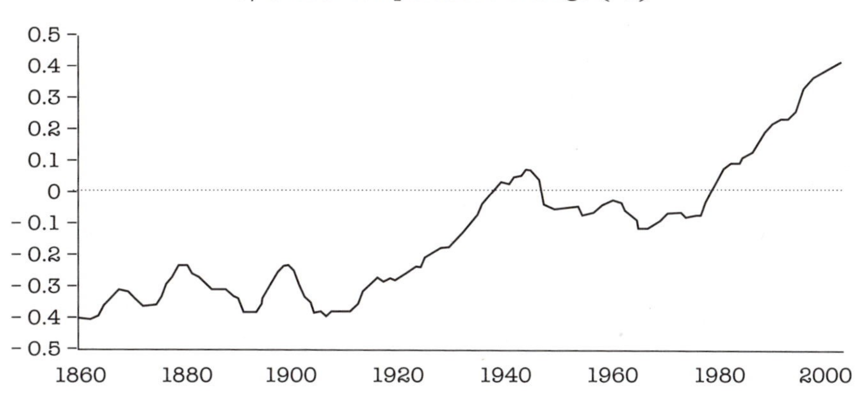

Source: Hadley Centre for Climate Prediction and Research, Met Office, UK, available at www.met-office.gov.uk.

weather like powerful hurricanes, tornadoes, and flooding, the expansion of tropical diseases, and deforestation are just some of the things we can expect. If we're not lucky, a recent report from the National Academy of Science warns, gradual global warming could couple with other human impacts on the environment and 'trip the switch' for a more abrupt and extreme climate shift.[16] Apparently small environmental changes can feed off each other, creating a complex feedback loop that accelerates and intensifies large-scale change and simply can't be stopped. Research over the past decade indicates that recent severe droughts and other regional climate changes may signal the onset of abrupt and dramatic climatic shifts that have not been seen on Earth since the last Ice Age.[17] If these scenarios prove to be even partially correct, we're in more trouble than even the most pessimistic estimates predict.

The recent science fiction film *The Day After Tomorrow* depicts massive hurricanes, floods, blizzards, tidal waves, tornadoes and a dramatic and sudden drop in temperature, as the next Ice Age appears literally overnight. While the film clearly exaggerates the pace of change, it may not be such an exaggeration of the ultimate consequences of global warming. The final effects of climate change may be just as devastating as the film depicts, but they will sneak up on us over decades rather than days.

Even cautious estimates put the most destructive effects of global warming only a couple of decades away. In fact, we can already see some of the early harbingers: global temperatures are rising far faster than the estimated rate at which ecosystems can adapt, and temperatures in some areas are rising at twice the global average.[18] Globally, the 1990s were the warmest decade since records have been kept. Record heat has been experienced in regions around the world. The majority of mountain glaciers have been melting.[19] At present rates, most will completely disappear by the end of the century. Glacier National Park in Montana is expected to lose its ice cover as early as 2030.[20] Recent resurgences and re-emergences of diseases are also linked to climate change.[21] The frequency and intensity of droughts have increased in many regions as well.

Given the amount of carbon dioxide already in the atmosphere, a significant amount of warming is unavoidable no matter what actions we take. And the effects of global warming can in turn intensify the warming trend, creating a kind of feedback loop. For example, the oceans now absorb perhaps 40 per cent of our carbon dioxide emissions from burning fossil fuels. As the marine ecosystem warms, it will likely absorb less carbon dioxide, thus increasing the warming. As evaporation increases with warming, the increased water vapor in the atmosphere may also add to the greenhouse effect.[22] Additionally, as arctic ice melts, the frozen soil releases carbon dioxide, intensifying the warming trend.[23] A similar feedback effect appears to be taking place in peat bogs in Europe and North America. 'The rate of acceleration suggests that we have disturbed something critical that controls the stability of the carbon cycle in our planet,' says research scientist Chris Freeman. 'On these trends, by the middle of the century, DOC [dissolved organic carbon] emissions from peat bogs and rivers could be as big a source of CO_2 to the atmosphere as burning fossil fuels.'[24] In short, the effects of global warming will help accelerate further global warming!

The destruction of forests will produce a similar cycle. Climate change is predicted to lead to a significant decline in tropical forests.[25] In fact, there is pretty clear evidence that climate change has already caused forest destruction in Europe and the US.[26] Forests absorb carbon to live and grow; and forests function to cool the planet by absorbing the sun's heat. So, as global warming attacks the earth's forests, their destruction will accelerate further warming.

No one can say for certain what all the consequences of global warming will be, but the results of extensive research give us a clear idea that they won't be good. The most serious impact will be linked to a rise in ocean levels. Even a slight increase in the temperature of the world's oceans causes an important rise in sea level. Water expands as it warms, which, when combined with the effects of melting ice caps and glaciers, may lead to a sea level rise of more than 19 inches over the next century.[27] Over the past century the sea has risen nearly 5 inches, but the rate of rise has doubled over the past few decades.[28] Because ocean temperature

1.6 / Rising sea levels (cm)

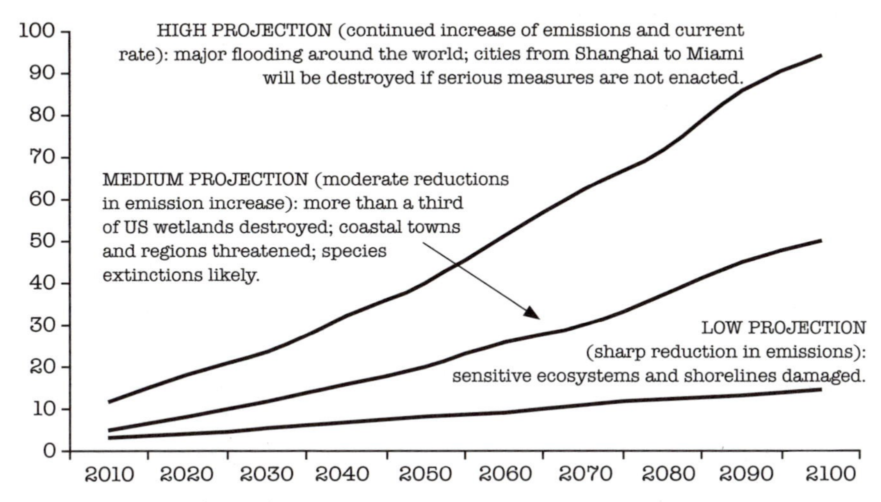

Source: Hadley Centre for Climate Prediction and Research, Met Office, UK, available at www.met-office.gov.uk.

rise lags behind the rise in atmospheric temperature, even if we stabilize greenhouse gas concentrations now, sea level rise would continue to occur for decades.[29]

The consequences in the United States will be severe. The Environmental Protection Agency has warned that a 4 foot rise in sea level along the Gulf and Atlantic coast is possible over the next century. Because a 1 foot rise in sea level would place more than 200 feet of shoreline under water in many areas, shore communities will be threatened across the east coast.[30] You may not only need to say goodbye to your favorite beach; you may need to say goodbye to your favorite beach town. By 2100, 10,000 square miles of the United States will no longer exist – that's an area the size of Massachusetts and Delaware combined![31] This might include portions of the nation's capital. Built on a former swamp, Washington DC is especially susceptible to flooding brought on by global climate change. Scientists in Florida, for example, have issued a clear warning: as the Florida Keys, Miami Beach, Fort Walton Beach, and many other tourist destinations face rising tides, the state's $47 billion tourist industry is threatened.[32] Higher temperatures and changing water cycles are likely to hurt the state's citrus industry as well. But Florida won't be alone. There will be plenty of pain to go around.

The impact on the earth's wildlife and forests is likely to be severe. Forests will be unable to adapt to changes in climate and rainfall, leading

1.7 / Estimates of the annual costs of global warming in the US

	Cost per year ($ billion)
Loss of land and dyke construction	7–9
Loss of water supplies	7–16
Losses in agriculture and forestry	9–21
Increased deaths	6–37
Increased power demands	8–10
Other factors	10–18
Overall cost	61–74

Source: John Houghton, *Global Warming: The Complete Briefing*, Cambridge: Cambridge University Press, 2004.

to increased vulnerability to disease and pests as well as stunted growth. Increased summer heat will increase the frequency and perhaps the severity of both forest fires and droughts. Because animals will be unable to migrate to cooler climes, this loss of habitats will cause an equally dramatic loss in species. Already the ranges and migration patterns for everything from polar bears to butterflies are changing as a result of the earlier arrival of spring. In many cases this has placed stress on wildlife reproduction.[33] Plants are blooming out of season. Scientists warn that such disruptions are likely to have a very upsetting effect on the delicate balance of ecosystems.[34] If, for example, a predator migrates north in an attempt to adapt to changes in water sources or forest cover, it may leave the prey on which it depends behind. This might be good for its former prey but as the prey's population grows, the plants or animals on which it feeds may become threatened. Fish populations in lakes and oceans around the world could be devastated as well.[35] The crash in Pacific salmon populations in the late 1990s is clearly linked to warmer ocean temperatures. Warming has also played a role in the decline in cod populations in the North Sea.[36] Coral reefs in more than thirty countries are dying due to rising ocean temperatures, threatening some of the most biologically diverse habitats in the world.[37] Fragile and ecologically rich wetlands, mangroves, marshes, and islands are being flooded by rising sea levels.[38] Two massive recent surveys reviewed over two hundred long-term studies of thousands of species, some of which encompassed over a century of change. Researchers tracked important changes in breeding behavior, blooming cycles, and migration patterns. Their findings are disturbing: we have already lost species as an effect of global warming.[39] Over the next fifty years, global warming will lead to the extinction of as many as 1 million species around the world.[40] That's about one-quarter of the world's land animals and plants.

A dramatic demonstration of the consequences of climate change on wildlife is being tragically imposed on polar bears. The world's largest carnivores on land, polar bears are a dramatic and beautiful demonstration of nature's capacity to adapt to various climates. With huge paws to help them grip ice and hunt seals, white fur for camouflage in the Arctic, black skin to absorb heat, and the capacity to carry lots of fat, the polar bear is perfectly adapted for life on the Arctic ice. Yet, if the current tends continue, polar bears may well disappear by the end of the century. As Arctic ice melts, polar bears, and other animals that make the Arctic their principal habitat, are in increasing trouble. Bears rely on their ability to put on plenty of fat in the winter to get through the summer of fasting when there is no ice to hunt on. So, with the Arctic ice breaking up earlier and earlier each year, polar bears are struggling to get by with a shorter eating season and a longer summer fast. A study in Hudson Bay has found that, as a result of shrinking ice, both body weight and reproductive rates of these magnificent creatures are clearly on the decline. Females and nursing young in particular are showing signs of distress. To make matters worse, climate-change-related stresses on the forests are threatening the habitat for polar bear summer dens. But it's not just bears: seals, nesting birds, and many other arctic animals are likely to pay the first – but not the last – price for global warming.

Yet it's not just wildlife that will pay the price. Even in the short term, the human costs of climate change will be considerable. Americans are likely to experience a rise in respiratory infections and mosquito-borne diseases such as malaria, dengue fever, and encephalitis.[41] Even increasing heat itself can have severe consequences: more than 700 people died in the 1995 Chicago heat wave; 15,000 died in France alone when a severe heatwave hit Europe in 2003. The World Meteorological Organization has warned that we can expect a dramatic rise in such severe weather as a result of climate change.[42] Already hurricanes, tornadoes, and other severe weather phenomena cause over $100 billion in damages and more than 40,000 deaths each year.[43] These numbers will rise. But this is only one of the effects. Additional side effects of global warming range from malaria to malnutrition and hit children particularly hard. A recent study by scientists at the World Health Organization and the London School of Hygiene and Tropical Medicine has already put the annual death toll due to global warming at 160,000.[44]

Severe weather events related to climate change are likely to cause the greatest suffering in the world's poorer countries. About 95 per cent of the 80,000 deaths that occur every year from natural disasters are in poor countries.[45] In 1998 flooding in China killed nearly 4,000 people. In the same year, Hurricane Mitch caused over 11,000 deaths and unimaginable human suffering in Central America; feeble homes and soil stripped of vegetation by an already impoverished population were swept away by massive floods and landslides. The following year, a cyclone in India caused well over 10,000 deaths and damaged nearly 2 million homes.[46] Two years later this region was hit by one of the worst droughts in history. Examples, unfortunately, are not in short supply.

Wealthy countries like the US may be able to put off the most destructive effects of global warming temporarily, but the cost will be very high. Dikes and costal barriers can be built to fight off the encroaching sea. For example, for an investment of $60 to $85 million, the destruction of Maryland's shoreline could be temporarily put off.[47] Farmers will need to spend millions adapting to changes in crop yields as currently arable lands grow unproductive with warming, and farming is forced further north. New patterns of plant diseases, pests, weeds, and changes in rainfall will add to these costs. We can expect food costs to increase substantially. It is likely that pressure to convert forest to croplands would increase as a result. Energy supplies could be threatened by changes in hydroelectric sources, and energy cost could rise substantially as a result. Swiss villages have already been forced to build earthen dams to protect their homes from rock and mud slides brought on by melting permafrost.[48] The direct financial cost of global warming will quickly run into the billions; the long-term indirect costs could dwarf that amount.

Some of these costs are already evident. In the first six years of the 1990s, insurance companies covered claims in excess of $50 billion as a result of climate-related catastrophes.[49] This was well above all previous expectations, and seems likely to rise dramatically. In fact, roughly 50 per cent of all catastrophe-related insurance claims since World War II have been paid out in the past decade.[50] The Global Commons Institute estimates that the damage due to climate change may reach $400 billion by 2012 and a whopping $20 trillion by 2050 – that's a 2 followed by thirteen zeros![51] Climate change could get costly.

The human costs will of course be incalculable. Bangladesh provides an extreme example. With 120 million people living in the country's delta region, a small sea-level rise could displace millions; and water levels could rise as much as 6 foot over the next century.[52] The people of Bangladesh do not have the resources to deal with such a disaster. Since sea-level

1.8 / Unequal impact (CO_2 emissions, per capita tonnes/year)

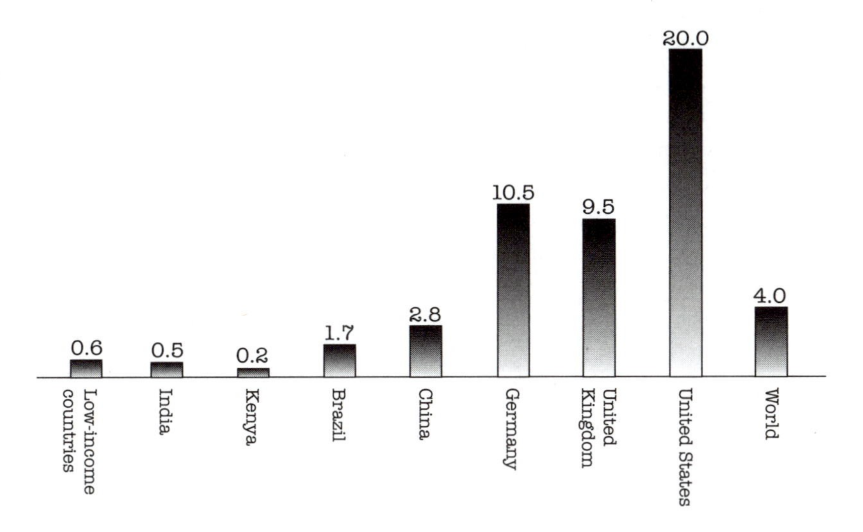

Source: *Human Development Report, 1999/2000*, United Nations Development Program, 2000, available at www.undp.org.

rises would likely devastate the country's primary industry, agriculture, the resources available to deal with the effects of climate change will decrease as climate change takes its toll. Many in the region now live at the edge of survival – they will die. There is no land to replace what will be lost; there is no place for the people of Bangladesh to go.

Bangladesh is not alone. Many regions in Africa and Asia will face similar catastrophes. In fact, more than half the world's population live in costal zones, where even a slight rise in ocean levels can have severe effects. Low-lying small islands have the most to lose from rising ocean levels. They will lose both land and fresh-water supplies to the rising sea. The cost of dealing with this crisis is far beyond the resources of the half-million residents of small islands in the Pacific and Indian Oceans. If we do nothing, these countries will quite literally disappear.

More generally, the dramatic weather changes brought on by global warming – hotter, dryer summers in some places, drenching rains in others – could make it impossible for farmers in the impoverished world to adapt. Severe water shortages could result from changes in water flow and quality linked to climate change. As a result, an additional 1 billion people will suffer from a shortage of fresh water.[53] The increase in vector-borne diseases, already a major cause of death in tropical countries, could devastate the millions who do not have access to adequate healthcare. In short, the serious deprivation of the world's disadvantaged majority will get worse fast; and the number of environmental refugees worldwide, already at an estimated 25 million, will grow quickly.[54]

According to the World Health Organization (WHO), 22 million more people are likely to suffer from hunger as a result of changes brought on by global warming.[55] Since 1.5 billon people depend on local agriculture for their daily subsistence, this may be a very low estimate. The prevalence of undernourishment around the world is already horrifying. Some 43 per cent of the population of sub-Saharan Africa suffer from undernourishment, and the number is growing. In South Asia, 16 per cent are undernourished; in Latin America it is 15 per cent, in the Near East, 12 per cent. In total, roughly 840 million people in the impoverished world suffer from undernourishment.[56] That's nearly four times the total population of the United States suffering from hunger around the world. What's more, an even larger number, a full quarter of the world's population, do not have easy access to fresh water.[57] Worsening these conditions with the effects of climate change is unconscionable.

It's an obscenity that the most destructive initial effects of climate change will be felt by the world's poorest, since they are the least responsible for this devastation. Carbon emissions per person in the United States are 6 tonnes per year. This is nearly ten times higher than the figure in Latin America and well over twenty times higher than the number for India.[58] In addition, much of the greenhouse gas produced in poorer countries is in the production of commodities to be exported to wealthy countries or consumed by the wealthy elite in the poor countries.

This global inequity is marked by the differences in the use of cars by the world's wealthy and the world's impoverished. Although automobiles are not the largest cause of greenhouse gas emissions, they are a major source, accounting for nearly 1 billion tonnes of carbon dioxide every year in the US alone.[59] The average American uses twenty-five times more energy for transportation than a person in a typical impoverished country.[60] Americans maintain six times the number of cars per person as the world average.[61] At the same time, average fuel efficiency in the United States has dropped dramatically over the past two decades, despite relatively stable rates of fuel efficiency in Europe and Japan.[62]

The rest of the world will not catch up. They can't. First, the wealth and consumption of the world's wealthy is only possible because of the poverty of the world's poor – the availability of cheap cars and cheap gas is dependent on the exploitation of cheap resources and labor around the world. This will be laid out in greater detail later, but put simply, the continued consumption by the rich requires a disempowered population

1.9 / Expected growth in motor vehicle use (billion miles traveled)

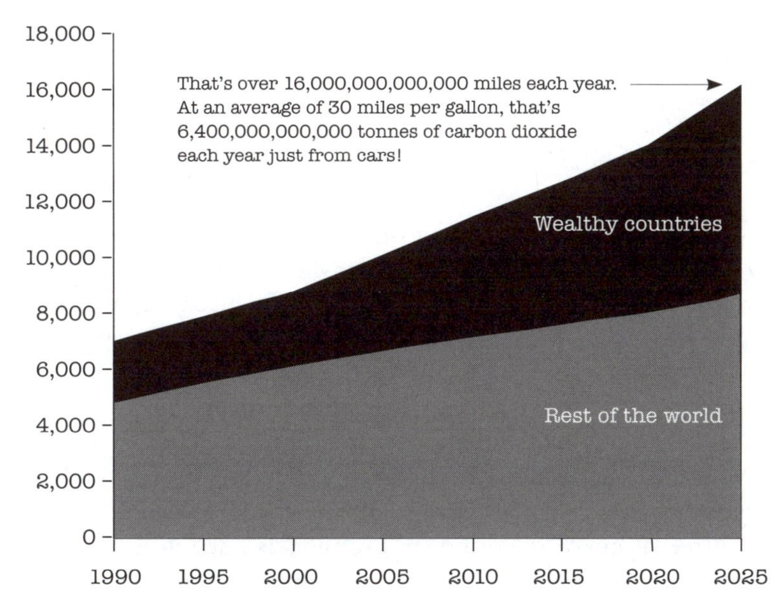

Source: OECD, *Motor Vehicle Pollution: Reduction Strategies beyond 2010*, Organization for Economic Cooperation and Development, Paris, 1995.

on whom the risks and waste may be dumped. Second, the earth's natural resources could not sustain American levels of consumption by all the world's people. If everyone on earth consumed the amount of resources and produced as much waste as the average American, we might not survive more than a few months.

Nevertheless, as the world's impoverished majority rightfully attempt to improve their lot and build a more tolerable existence, their increased consumption will accelerate climate change. The use of motor vehicles in the rest of the world is projected to grow. By 2030, the number of miles traveled in the rest of the world should match the amount in the world's wealthiest countries.[63] Of course, since the citizens of these wealthy countries comprise only some 15 per cent of the world's population, this will still reflect a gross imbalance. Even so, figure 1.9 shows, it will mean billions of additional tonnes of carbon in the atmosphere every year.

All of these devastating developments apply only to the relatively short-term picture, of course. In the long term the perspective is considerably bleaker! Despite their high cost, measures taken by the world's wealthy to mitigate the effects of climate change will provide only temporary protection. If we continue with business as usual, the rising waters will eventually spill over our dikes, and not only small islands will be swamped. As the pace of global warming continues, even the wealthy will have to face the dire consequences.

The air we breathe

Global warming isn't the only effect of the increasing damage that we've done to the atmosphere. In the late 1970s we discovered that our use of chlorofluorocarbons (CFCs) for decades in industry, agriculture, and households had severely damaged the earth's upper atmosphere. The chlorine and bromine released as these chemicals break down in sunlight were destroying the upper ozone layer that protects all life on Earth from excessive ultraviolet radiation from the sun. The good news is that international negotiations have led to a dramatic drop in the use of these chemicals; and that the damage may not be irreversible. However, since one free atom of chlorine or bromine is capable of destroying as many as 100,000 ozone molecules, the damage continues to worsen. Moreover, since most of the CFCs that have been released are still in the atmosphere, and the lifetime of ozone-depleting chemicals is known to be up to one hundred years, full recovery will not take place until many decades after the release of these chemicals has stopped.[64] And, although the production of ozone-depleting chemicals has declined significantly, it has not stopped. CFCs, hydrochlorofluorocarbons (HCFCs), and other ozone-depleting substances are still being produced and used in the US and around the world. And leakage from existing sources, such as old refrigerators, air conditioners, and Styrofoam products, continues to be a major source of these destructive chemicals. If all goes well, the ozone layer might be restored to its 1980 level by 2050 – but all may not be going well.[65]

The depletion of the ozone layer will likely have severe public health consequences around the world. Among the likely effects is a greater susceptibility to infectious diseases, increased incidence of cataracts, and skin cancer. EPA estimates indicate that in the US alone the decline in the ozone layer is responsible for more than 12,000 additional cases of potentially fatal skin cancer each year.[66] Although skin cancer rates are most likely to rise among fair-skinned Europeans, Americans, and Australians, the people of Asia, Africa, and Latin America will share the pain. In parts of the world where malnutrition and disease are already a major problem, increased susceptibility to infectious diseases is particularly

1.10 / The ozone hole continues to grow (million km²)

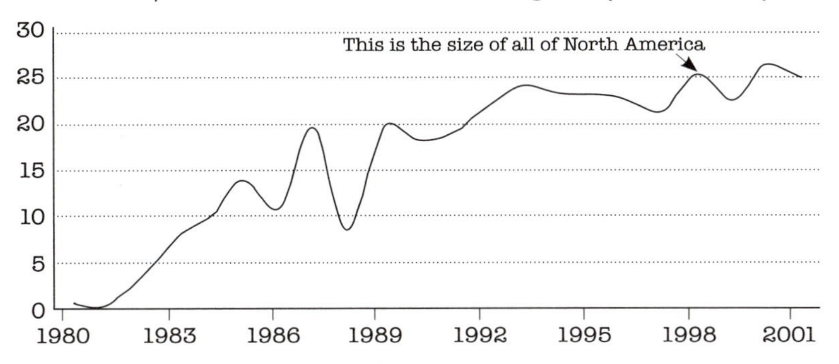

Source: US EPA, *The Size and Depth of the Ozone Hole*, www.epa.gov/ozone/science/hole/size.html.

bad news. Since the late 1990s, the hole over the Antarctic has extended over southern Chile; and even a few minutes of exposure to the sun could lead to irreversible damage to skin and eyes.[67] While the wealthy might be able to rearrange their day to avoid the sun, the impoverished do not have that luxury.

Greater UV exposure can also damage plant and wildlife. Leaf damage, loss of photosynthesis, and stunted growth have been identified in plants exposed to enhanced UV. Wildlife can also be affected; scientists have already noted rapid declines in reptile populations on every continent as a result of increased UV exposure.[68] Of course the most damaging effect may be indirect; as ecosystems are exposed to higher UV, the damage to food sources or other aspects of the ecological balance could have serious consequences.

The world's oceans may pay the highest price for ozone depletion. High doses of ultraviolet radiation kill the phytoplankton and krill that serve as food for much sea life. So, by allowing more ultraviolet radiation to reach the ocean's surface, the destruction of the ozone layer could eliminate large numbers of ocean species. It doesn't end there: since phytoplankton are a major absorbant of carbon dioxide, destroying them will reduce the earth's capacity to absorb greenhouse gases, thus intensifying climate change.

At the same time, global warming can also worsen the depletion of the ozone layer.[69] Greenhouse gases warm the air near the earth's surface but cool the air at higher altitudes. Since the chemical reaction that results in the destruction of ozone is increased at colder temperatures (explaining why the ozone hole first appeared over the South Pole), this cooling worsens the deterioration of the ozone layer. Scientists have discovered that the severity and duration of the ozone hole have already been made worse by the effects of global warming; and climate change may undo some of the benefit we've achieved by banning CFCs.[70]

Anyone who lives near a major city knows that the costs of air pollution are not limited to the destruction of the ozone layer or climate change. That thick gray-green haze that hangs over cities around the world is a ready reminder of the trouble we're in. Industrial emissions of nitrogen oxide and other contaminants react in sunlight to create this low-level ozone. Unfortunately, this ozone does not rise in the atmosphere to replenish the stratospheric ozone layer. Instead it stays near the ground, in the troposphere, where it creates smog, contributing to human and animal illness and ecological damage. The immediate health consequences of breathing polluted air are severe and absolutely clear; they can be measured in rising hospital admissions, increasing disease rates, and mounting deaths.[71] Air pollution can impede the brain's proper functions and permanently damage the human body. Bronchitis, emphysema, colds, asthma, and other health problems can be considerably worsened by the sulfur dioxide in polluted air; while the high levels of nitrogen oxide in smog can lead to or intensify influenza, pneumonia and bronchitis. Excess emissions from older, coal-burning power plants alone lead to 18,700 deaths each year.[72] More generally, air pollution significantly shortens and degrades the lives of all Americans who breathe contaminated air. In Los Angeles, nearly 6,000 deaths each year are attributable to air pollution; in the New York City area, each year 4,000 deaths are attributable to air pollution; in Chicago, 3,500, in Philadelphia, 2,500, and the list goes on.[73] Overall, as many as 60,000 to 120,000 deaths annually in the United States can be attributed to the effects of air pollution.[74] In all, 243 counties in the US, home to nearly 100 million people, failed to meet federal air standards in 2004.[75] The annual cost of healthcare, lost productivity, and restricted activity days attributable to air pollution are measured in the tens of billions of dollars. Around the world, perhaps as many as 700,000 deaths are brought about by air pollution each year.[76]

The very young, the elderly, the poor and the otherwise marginalized are most harmed. Severe asthma affects people of color and the poor disproportionately, leaving them especially vulnerable to the effects of air pollution.[77] More than 51 million Americans live in areas where smog levels exceed safe standards; a disproportionate number of them are black, Hispanic, or poor.[78] We know that black, Hispanic, and impoverished children have for decades been poisoned by lead, ozone, dust, and carbon monoxide at rates that far exceed the American average.[79]

1.11 / Environmental racism
(air quality non-attainment areas by pollutant and ethnicity, %)

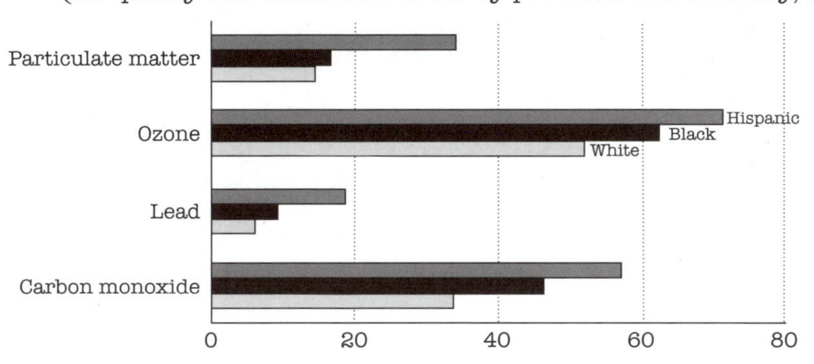

Source: EPA, *Environmental Equity: Reducing Risk for All Communities*, Environmental Protection Agency, 1992, available at www.epa.gov.

Globally, once again, the impoverished and marginalized are the most affected. Over 1 billion people, or one out of six people in the world, live in air that is considered unfit to breath.[80] In Mexico City, air pollution exceeds world health standards 300 days each year.[81] Nearly 3 million people worldwide die each year from the effects of air pollution.[82] Over two-thirds of these are impoverished women who inhale smoke and fumes from fuelwood and dung in inadequate and unsafe homes.[83] Although lead emissions are recognized as a major cause of irreversible damage to the central nervous system and have been severely reduced in wealthy countries, in Latin America roughly 15 million infants under the age of 2 are at risk of brain injury from high lead emissions.[84]

So the poor take the hardest hit. Yet air pollution is a global problem, and no region or person will be spared its overall effects. Atmospheric scientists have discovered that Asian smog can travel thousands of miles across the Pacific Ocean and contribute to ground-level ozone in the United States.[85] In fact, increased air pollution in China can make it more difficult for US states to meet clean air standards.[86] No place on earth is immune; toxic industrial emissions have been found in the air over the Arctic Ocean.[87] So, without a global clean-up, local and national smog controls are unlikely to save us.[88] To make matters worse, the effects of global warming will make air pollutants slower to disperse and will make us more susceptible to their harmful health effects.[89]

Still, that's not the end of our capacity to degrade our atmosphere. Since the late 1960s, we've known that rain and fog across the US and

1.12 / The pH scale

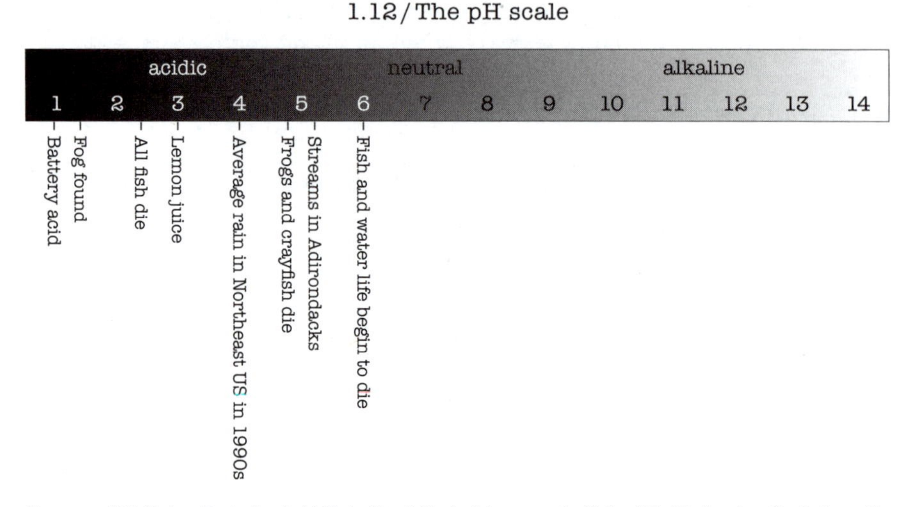

Ssource: C.T. Driscoll et al., *Acid Rain Revisited: Advances in Scientific Understanding since the Passage of the 1970 and 1990 Clean Air Act Amendments,* Hubbard Brook Research Foundation, Hanover NH, 2001, available at www.hbrook.sr.unh.edu.

Canada were becoming more acidic.[90] This acid rain primarily results from the large amounts of nitrogen oxide and sulfur that are pumped into the atmosphere by the smelting of sulfide ores and the combustion of fossil fuels. In the US, the single biggest source is outdated coal-fired power plants. When these pollutants combine with the water in air, they become two of the strongest acids known (sulfuric and nitric acids). This acidic water vapor can be carried hundreds of miles from the pollution source. Although unpolluted rain is naturally slightly acidic, rain and snow in many regions of the world have become up to thirty times more acidic than the most acidic rain occurring in unpolluted conditions. Rain has been recorded in parts of the US that is more acidic than lemon juice or vinegar! In parts of the world, fog has been recorded with an acidity approaching that of battery acid.[91]

It is absolutely clear that acid rain decimates forests close to the source of the pollution.[92] The emissions from the ore-smelting operation in Sudbury, Canada, provides a clear example. The International Nickel Company's operation there has long been the world's greatest single source of sulfur dioxide. The plant built the largest smokestack ever constructed (1,250 foot) in order to minimize the local impact of the 2,500 tonnes of sulfur dioxide released into the atmosphere every day. The effect has been to obliterate several hundred square miles of habitat downwind from the smelter. All vegetation is gone. The lakes in the area are completely without life. The soil is barren. This is an extreme example, but not a lone one.

Perhaps even more disturbing, acid rain and smog can cause very severe damage to forests at greater range.[93] The death of trees in the Sierra Nevada mountains can be linked to growing air pollution in the San Joaquin valley, well over 100 miles away.[94] Trees in forests stretching from Ohio to Tennessee are dying at three times their natural rate, attacked by airborne toxins from faraway power plants and factories.[95] In remote areas of the Rocky Mountains, emissions from plants hundreds of miles away have led to damage to fragile watersheds and pristine mountain meadows. Acid leaches minerals from the ground; and acid rain or fog damages leaves and therefore the tree's capacity to breathe. Overall, it is apparent that acid rain increases the stress on soils and forests, making them less capable of resisting other threats.[96] So forests weakened by acid rain are less able to withstand the stress of storms, insects, disease, water shortage, or severe winters.[97] Forests in Canada, the Northeastern US, Northern Europe, and the Western US have been severely affected.

The effects of acid rain on freshwater sources is also dramatic. Up to 20 per cent of the lakes in the Adirondack Mountains are no longer habitable for sensitive fish species.[98] Polluted rain also leads to the dissolution of mercury into freshwater and the fish that live there. The people who eat the contaminated fish suffer a damaged nervous system, loss of motor skills, weakening muscles, and eventual paralysis. Across the United States, fish in roughly 50,000 lakes, streams, and ponds contain levels of mercury that are harmful to humans.[99] Excess acidity can also raise levels of arsenic, cadmium, aluminum, and selenium in groundwater.

Acidification can reduce the number of fish species in a lake from several hundred to only a few. Even the handful of species that can live in acidified water accumulate mercury in their body tissue, causing the birds and mammals that live off those fish to suffer mercury poisoning. In the North East wilderness each spring, rivers and lakes experience an 'acid shock' as the snow pack melts, jolting waterways with a huge jump in acidity at the start of the growing season, just when plants and animals are most vulnerable.[100] As more and more lakes grow too acidic to support their full spectrum of life, so birds, fish, frogs, crayfish, otters, and other once-thriving inhabitants grow increasingly rare. As we will see in later chapters, efforts to regulate acidifying emissions have proven clearly inadequate as the strain on our forests, rivers, and lakes increases.[101] In fact, current emissions are roughly five times the level that would allow even a partial recovery of our lakes, fish, and trees.[102]

At this point, it should be no surprise that acid rain can combine with the effects of climate change and ozone depletion to intensify the stress on forests and lakes.[103] In fact, large parts of the world will suffer the combined effects of acidification and climate change in the coming decades.[104] Researchers have concluded that our lakes are under a 'three-pronged attack' as increased exposure to ultraviolet radiation due to the depletion of the ozone layer combines with climate change and acid rain to destroy their capacity to support life.[105]

To complicate matters further, these pollutants lead to the accumulation of dust, or aerosol particles, in the upper atmosphere that can partly mask the effects of greenhouse gases.[106] These tiny particles reflect the sun's heat, and have a cooling effect locally. But this effect is less powerful than, and doesn't last as long as, the warming effect of greenhouse gases (weeks as opposed to decades). This means that as we reduce emissions to deal with climate change, the cooling effect of aerosol particles will end faster than the greenhouse effect, leading to accelerated warming.

The global damage to the atmosphere is likely to mount. For example, the geographic scope of acidification has so far been relatively limited because the industries associated with the production of acidifying pollutants have been focused in a relatively few countries of the world. As regions in Asia, South America, and Southern Africa increase their industrial production, they will increase the production of acid rain and other pollutants.[107] If the world's impoverished countries follow the example of the world's wealthy countries, the global environmental damage outlined thus far will be dwarfed by the catastrophe yet to come.

Ecocide

Deforestation is not a new phenomenon, but the rate of forest destruction has increased substantially over the past half century and become a central element in our war on nature.[108] The most significant cause of deforestation is not acid rain but intentional destruction of forests for timber, grazing land for cattle, and land for agriculture. Tractors, chainsaws, and other modern 'conveniences' have allowed for the destruction of some of the earth's last great forest frontiers at unparalleled rates. Roughly 34 million acres of forests are lost each year.[109] Imagine a forest one hundred miles wide and five hundred miles long – an area greater than the size of New York state – lost every year.

Even this staggering rate of global deforestation underestemates the depth of the problem. Official figures for forest loss do not reveal the fact that ecologically rich and irreplaceable virgin forests are being destroyed and replaced by corporate monoculture – large-scale tree plantations designed for maximum profit.[110] These are not forests, they're pulp fiber factories,

1.13 / Global forest change, 1980–1995 (%)

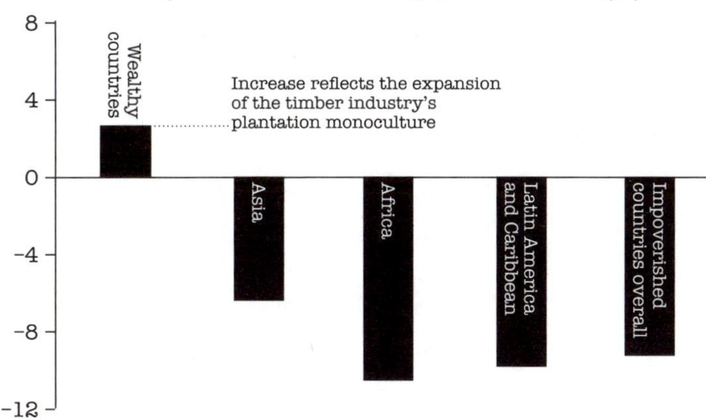

Source: Emily Matthews, Richard Payne, Mark Rohweder, and Siobhan Murray, *Pilot Analysis of Global Ecosystems: Forest Ecosystems*, Research Report, World Resources Institute, Washington DC, 2000 p. 17.

row after row of evenly spaced identical trees, bred for rapid growth, sprayed with herbicide and mechanically harvested at regular intervals. Wildlife is scarce on these plantations and the normal functions of a forest ecosystem do not exist. The effects of increased soil erosion, water contamination, and biodiversity loss extend well beyond the plantation itself. This process can be neither sustainable nor a stable source of economic development. Yet, with the demands of paper manufacturers growing, these industrial plantations are expanding around the globe, creating environmental costs that are still largely unrecognized.[111]

The immediate consequences of deforestation can be severe. Forests play a crucial role in the natural water cycle, for example. Rain falls in mountain forests, gradually seeps through sand, soil and rock, and emerges from this natural filtration system as clear river and lake water. When the trees are removed, the water simply runs off mountainsides in sheets, taking topsoil and rock with it, and leading to soil erosion, flooding, landslides, and the end of any natural purification system. Deforestation has led to massive floods in China, India, Malaysia, and Latin America, costing tens of thousands of lives and billions of dollars in damage.[112] As global climate change takes its course, we cannot afford to make ourselves more vulnerable to flooding. In addition, habitat destruction leads to a decline in predators of disease-carrying animals. As these prey overexpand, they contribute to the spread of sickness. In fact, several epidemics can be linked to deforestation.[113] Remember that global climate change itself is also leading to the expansion of diseases. The combined effect is likely to be significant.

1.14 / Number of species documented as extinct or threatened in 2002

	Extinct*	Endangered	Vulnerable	At risk‡	Total
Mammals	77	520	617	936	2,150
Birds	132	508	684	814	2,138
Fish	91	299	442	432	1,264
Reptiles	23	134	159	137	453
Amphibians	7	67	90	99	263
Invertebrates	380	758	1,174	957	3,269
Plants	92	2,337	3,377	1,344	7,150
Total	802	4,623	6,843	4,719	16,687

* Includes species documented as extinct and extinct in the wild. Because a species must be missing for decades before it is listed as extinct, this category is highly conservative.
‡ Includes near threatened, conservation dependent, and indeterminate.

Source: *Red List of Threatened Species*, World Conservation Union (IUCN), Gland, 2004, available at www.redlist.org.

We are only just beginning to understand the value and vulnerability of the biological diversity we are destroying through deforestation. We do not know exactly how many species there are on earth. Of the estimated 9 million, we have identified roughly 1.5 million. We are particularly ignorant of the species of Africa, Latin America, and Asia, where fewer scientists are present to study them. And the effects of even apparently moderate deforestation are severe for wildlife. Researchers know that the number of species supported by a particular habitat drops much faster than the reduction of the size of the habitat itself. So, if a forest of 100 square miles is reduced to 10 square miles, we can not only expect a dramatic drop in the actual number of plants and animals in that area, but we can also expect the variety of species in the remaining forest to be halved.[114] As we destroy these forests, we quite literally don't know what we are losing.

Although a certain degree of extinction is part of the natural process, the rate at which species are threatened is increasing dramatically. The disappearance of 626 animals has been clearly documented; and nearly 10,000 animal species are now considered to be threatened with extinction.[115] The National Audubon Society reports that more than one-quarter of America's birds are in decline or in trouble.[116]

Documented rates are only the tip of the iceberg, however. Modest estimates put the global rate of extinction at one species per day.[117] Noted biologist Edward O. Wilson estimates that the destruction of tropical

1.15 / Area of tropical forests, 1800–2000 (billion acres)

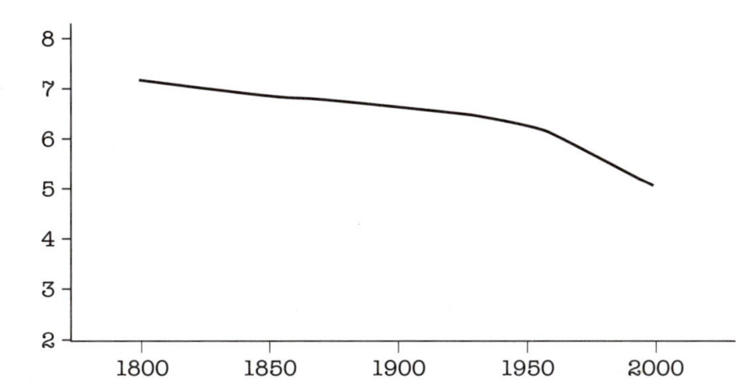

Source: Dhirendra K. Vajpeyi, 'Introduction,' in Dhirendra K. Vajpeyi (ed.), *Deforestation, Environment, and Sustainable Development*, Praeger, Westport, 2001, p. 3.

rainforests alone leads to the extinction of 27,000 species each year – that's 74 species per day.[118] The World Wide Fund for Nature (WWF) estimates that between 1970 and 2000, the population of the earth's forest, marine, and freshwater species has decreased a full 37 per cent.[119] Moreover, as global warming continues to change the climate more rapidly than species can adapt, we can expect these numbers to increase.[120]

The most noted consequence of deforestation is more directly tied to global warming. One acre of rainforest holds about 200 tonnes of carbon.[121] When the forest is destroyed, that carbon is released into the atmosphere as carbon dioxide. Indeed, the cutting down and burning of forests account for roughly a fifth of all carbon dioxide emissions.[122] As a result, globally, forests are now a net source of carbon![123] So, our dramatic and accelerating destruction of the earth's forests intensifies the horrifying effects of the greenhouse gases that we continue to pump into the atmosphere.

Tropical rainforests are particularly important to our effort to preserve the earth's ecological balance. The cooling effect for the earth's atmosphere provided by the Amazon rainforest is enormous. More than this, while they cover only 6 per cent of the earth's surface, the rich habitats of the tropical forests contain an estimated 50 to 80 per cent of all species; and these species are less able to adapt to changing climate and stress than species in North America or Europe.[124] Yet, in the time it takes you to read this sentence, 7 more acres of the Amazon will have been destroyed.

A loss of this rich rainforest biodiversity will have a significant impact on the future course of evolution. Rainforests contain the ancestors of many of the earth's plants; so, with the loss of the forest, the gene pool for future evolution is thinned.[125] Medicines for leukemia, heart ailments, hypertension, arthritis, birth control and Hodgkin's disease are derived from tropical plants.[126] Some 70 per cent of the plants known to fight cancer are tropical.[127] Plants from the Peruvian rainforests could help doctors fight tuberculosis, a killer of 3 million people worldwide every year.[128] According to the United Nations Environmental Program, we lose the possibility of one major drug every two years due to plant species extinction.[129]

Humans have always exploited forests, and deforestation is not new, as the timber and ranching industry often points out. But the current rate of tropical forest destruction and loss of biodiversity has no parallel in the last 66 million years. If the current rate of destruction continues, virtually the entire Amazon and all other tropical forests will be gone by about 2050. Well over half of the rainforest in Guatemala has been destroyed over the past three decades. At current rates, rainforests in the Philippines will be entirely gone in thirty years. Those in Malaysia will be gone in about fifty years. A short time later virtually all the rainforests of Central America will be gone.[130]

The devastation is often greater than the rates of destruction would imply, since bulldozers, road graders, log skidders and other machinery take their toll even on land that is not entirely cleared. In Brazil, for example, the amount of forest affected by deforestation is nearly three times the area that has actually been cut down.[131] Deforestation not only causes soil degradation and flooding; the silt that runs to rivers and streams from cleared lands kills fish, destroying a local food source as well as a crucial part of the ecosystem. And water scarcity, already at a critical level in many impoverished countries, grows more severe as the natural water cycle is destroyed. In Costa Rica, the destruction of lowland forests has reduced the life-enriching fog over the forest and placed the famed cloud forests on the mountain tops at risk.[132] Imagine a cloud forest with no clouds.

In addition to all these practical, objective reasons for concern, we should recognize the intrinsic, sacred value of the forests we are destroying. As uniquely rich habitats for some of the world's most beautiful plants and animals, tropical forests have an essential value which cannot be measured in dollars. If we allow them to die, we lessen our own existence. As we make the earth more barren, our lives and our collective future lose value.

The largest cause of global deforestation is the pillaging of forests in the world's poorest countries to feed an ever-growing demand for wood products in the world's wealthy countries. Every year, the wealthy countries of the world consume roughly 200 million tonnes of paper and 14 billion square feet of sawn wood and wood-based panels – that's enough to cover the entire states of California, Oregon, and Washington with wood panel![133] These figures are sharply on the rise. The massive consumption of even the simplest wood item often implies tremendous environmental cost. For example, chopsticks in Japan are made from nonrenewable tropical woods, not because this wood is superior but because it can be plundered from desperately poor countries at a low price. The chopsticks consumed in Japan alone could form a mountain the size of Mount Fuji every year.[134]

The world's poorest countries have opened up their forests to exploitation by international timber industries in the often desperate effort to raise money for development projects or to pay debt. While one might at least hope that the resulting destruction of the world's forests could allow the world's poor to improve their lot, the opposite is often the case. The cost of the roads, ports, and other facilities built by poor countries to woo multinational timber industries frequently exceeds the small amount of money skimmed by impoverished governments from the profits of Western timber industries.

In Central and South America, a leading cause of deforestation is the beef industry. Large landowners clear vast tracts of forest to provide pastures for the raising of cattle. Requiring four times the amount of land and water as the production of an equally nutritious amount of grain, the production of beef is notoriously environmentally unfriendly. A silver lining might be present if the beef industry led to more food for the local population; however, most of the beef production from grazing is exported to hamburger chains in wealthy countries. In fact, while beef production in Central America tripled from 1950s though the 1970s, beef consumption in the region actually fell. Three-quarters of all Central American beef exports are headed to the US.

Cattle grazing leaves the land decimated, stripped of nutrients, and unable to recover. The nutrients in a rainforest are contained in the plants not the soil. Contrary to popular perception, tropical rainforest soils are highly fragile and nutrient-poor. In the cutting and burning of these forests, the ash is mixed with the soil and makes it temporarily nutrient-rich. But the land remains suitable for agriculture or grazing for only a few years before it is depleted and the cattle rancher moves on to new areas of virgin forest. Cattle grazing for the international market thus devours increasing amounts of forest simply to maintain a constant level of productivity. Nearly all the ranches established in the Amazon region prior to the mid-1970s had been abandoned in little over a decade. The pace of destruction is often dramatic, with several square miles being 'slashed and burned' at a time.

Small-scale farmers also place demands on forests; though the potential for sustainable land use is far greater.[135] Typically, poor farmers chop down a few acres to raise crops for food and to sell. Again, because the rainforest soils are infertile and highly fragile, farmers are often forced to abandon land after a few years, moving on to clear new areas of the forest. Typically, the land can recover from this kind of small-scale agriculture in less than fifty years.[136] In contrast, large-scale agriculture for export, such as banana growing, uses large amounts of pesticides and fertilizers and leaves the land decimated indefinitely. In short, local populations struggle to survive while large corporations produce luxury food for the wealthy, destroying the earth as they go.

The livelihood of Brazil's rubber tappers serves as a powerful example of the millions of people around the world who depend on the forest for their livelihood and are being pushed aside. The tappers, or *seringueiros*, had learned how to live in balance with the forest from the ancestral traditions of indigenous peoples. When the government began efforts to promote colonization programs and investment from industries in southern Brazil to 'develop' the region and reap profits from the forest, the pressures on these peasant people and the forest grew dire. Large landowners purchased millions of acres of forest with the government's help, clearing the land for logging, industrial agriculture and cattle raising. Terror tactics were used to remove the *seringueiros* and indigenous peoples of Amazonia. When simple intimidation failed, houses were burned, cattle were killed, wives and children were threatened or abused, and resisters were assassinated. In 1988, wealthy landowners murdered Chico Mendes after he led a well-known resistance movement by local *seringueiros*. Although Mendes's murder turned the world's attention to the crisis, almost two decades after his death the forest is under greater threat than ever. Across Latin America, Africa, and Asia, indigenous communities are being driven from their ancestral homes, deprived of their livelihood, brutally treated, exploited, and murdered in a rampage to feed the desire for wood and beef by the world's wealthy.

Toxic planet

Environmentalism is too often associated only with birdwatchers, Sierra Club hikers, and animal lovers. Yet the ensuing crisis threatens every person on earth. My fate and yours are intimately entwined with the survival of the earth's other species. Certainly our cultural and emotional well-being is dependent on the richness of the earth's plants and animals. When we destroy them, we destroy a part of ourselves. However, more than this, by tearing apart the web of life we not only lessen our own existence, we endanger our physical survival. The destruction of even small, apparently unimportant species can have devastating consequences for the entire system and each of the species that depend on it. In fact, scientists often do not know the exact ecological role or function of a species. Our understanding of how the pieces fit together is still pretty sketchy. Some species are food for others; some are necessary for the survival or reproduction of others, some serve as host for others. What is clear is that no species – not even humans – escapes the complex interdependency of all living beings. Biodiversity isn't just about the survival of a plant, a newt, or a spotted owl, it is in the end about your own survival.

By poisoning the web of life on which we depend, we poison ourselves. A variety of what are called persistent environmental toxins have accumulated in the food chain through a process called biomagnification. This means that the concentration of toxins increases with each step in the food chain. So, while the toxins in plants may be small, they will be dramatically increased in the insects that feed on those plants. They will again increase many times over in the small fish that feed on those insects, and again in the large fish that feed on those small fish. In fact, the concentration of pollutants in water may be increased as much as 700,000 times in the bodies of animals living in that water. So, heavy metals, deadly PCBs, cyanide, selenium, and other dangerous contaminants build as they travel up the food chain. Human beings are at the end of that chain.

So the pesticides and heavy metals pumped into our water by industries and mining not only kill animals from insects to seals, they also increase toxin levels in people. The blood and urine of Americans are contaminated with more than a dozen pesticides.[137] The US geological survey tested 139 streams across thirty states in 2000 and found contaminants in a full 80 per cent.[138] Another recent study of thousands of fish from sites in the Atlantic region found that every tested location displayed at least one contaminant; a full third exceeded safe levels for four different contaminants.[139] Scores of pharmaceuticals are routinely found in our rivers and lakes, including antibiotics, antidepressants such as Prozac, pain relievers, and estrogen from birth control pills. Whether they make their way through a human body and are flushed into wastewater or are discharged without proper treatment from manufacturers, these compounds are ending up in our waterways at dangerous levels. The results can be male fish that are 'feminized,' growing eggs in their testes, other defects in fish and amphibian reproductive systems, and changing sex ratios.[140] The effects on humans simply aren't yet known.

No one is safe. A few years ago, researchers who were examining contaminants in breast milk in Canadian women sought samples of breast milk from Inuit women as a comparative sample. The idea was to get an example of pure human breast milk. So they sought out the Inuit people, living above the Arctic circle in what was they believed was a pristine, uncontaminated landscape. They were shocked to discover that the presence of mercury and PCBs in the breast milk of these women was in high enough concentrations to be considered toxic waste! The poisons emitting from power plants and industries thousands of miles away had accumulated in the plants and animals of the Arctic. Diets that had sustained their ancestors for generations were now poisoning these indigenous women and leading to serious brain and neurological damage in their children. In a similar discovery, researchers at the University of Alberta found toxic substances such as DDT and PCBs in what were thought to be pristine glaciers in the remote high peaks of the Canadian Rocky Mountains.[141] Distant areas of the Amazon, once thought to be too remote to be affected by human activity, now show signs of damage from the excess carbon dioxide in the atmosphere.[142] The upshot is that no community, no landscape, no individual is unaffected by the poisons we pump into our environment.

We have made the earth our toxic waste dump, pumping more and more toxic compounds into our lakes, valleys, fields, and meadows. Barrels of toxic waste appear in canyons; toxic sludge builds up in hidden meadows; poisonous pollutants are discovered in our playgrounds, schools, and backyards. Every year, a million tonnes of hazardous waste are produced; and, according to the federal government, a full 90 per cent of toxic waste is disposed of improperly. About 90 per cent of this waste is produced in wealthy, industrialized countries.[143] In the US alone, there are between 400,000 and 600,000 contaminated sites that must be dealt with.[144] Dioxins, highly toxic industrial by-products produced by paper mills, hazardous waste incineration, and the manufacturing of plastics and pesticides, are linked with cancer, birth defects, learning disabilities, infertility, immune dysfunction, reduced mental capabilities, and hyperactive disorders in children. With one gram of dioxin enough to exceed the acceptable daily intake for more than 40 million people for one year, the US Environmental Protection Agency reports that 99,814 grams of dioxin were released into the environment by chemical manufacturing and processing facilities in 2000.[145] Runoff from farm chemicals can be just as deadly. A recent study found that the drinking water of over one-quarter of the residents of Iowa contained traces of one or more pesticides.[146] Over half the wells in the US are polluted with nitrates or pesticides.[147]

Oceans, rivers, and lakes have become convenient sewers for the industries that feed our insatiable desire to consume. Sewage and other pollutants have increased nitrogen levels in seas and lakes. The resulting algae that feed on the nitrogen literally suck the oxygen from the water, suffocating the surrounding life. One dramatic example of this effect is the so-called 'dead zone' in the Gulf of Mexico. Commercial fertilizers, agricultural and municipal waste, and other forms of pollution from the Corn Belt states feed into the Mississippi river. This flow includes nearly 7 million tonnes of fertilizer every year. As these pollutants flow into the Gulf, they seed huge algae and phytoplankton blooms. These blooms suck the oxygen from the water as they grow, die, sink to the ocean bottom, and decompose. Oxygen-dependent sea life in a 8,500 square mile zone is killed. Each year, this dead zone grows and makes a dramatic impact on biodiversity, not to mention commercial fishing, in the Gulf. Such dead zones may be the greatest global threat to ocean life. These algae blooms appear around the globe, and can measure up to 43,500 square miles. They have doubled in size about every decade since the 1960s.[148]

Every year, roughly twenty million barrels of oil enter the earth's oceans from land runoff or spills, devastating local sea life. In 2000, there were 11,270 days of beach closings and advisories in the United States alone.[149] Some seals, dolphins, and whales are so contaminated with our chemical waste that they could qualify as toxic waste![150] Floating debris washes up on remote shores from Alaska to Antarctica. Plastic bottles are regularly seen floating in the open ocean, hundreds of miles from shore. These plastics are not biodegradable: although they do break down into smaller and smaller pieces, they remain as dust in our oceans and have untold effects on the ecosystem.

If we do not dump waste into our rivers and oceans, we try to bury it. The amount of waste we produce is staggering. The US alone produces 530,000 tonnes of waste each day. Despite what is sometimes claimed, this waste is essentially not biodegradable. Forty-year-old newspapers can be dug up and read. Even organic material like yard waste and food degrades very slowly.[151] Plastics simply do not degrade; and 6 million tonnes of plastics are added to American landfills each year.[152] Americans throw away 2.5 million plastic bottles every hour![153] Only a very tiny percentage of plastic products, perhaps 2 per cent, are recycled.[154]

And things are getting worse. The proportion of recyclable glass and metal is actually decreasing. Some 40 per cent of household waste is unnecessary packaging, and this proportion is also on the rise. Computers present a clear example of a growing threat. With an increasingly short useful life, computers are a growing addition to our waste stream – and a serious toxic hazard. The National Safety Council estimates that discarding our roughly 315 million old computers over the next few years alone would contribute a total of about 1 billion pounds of lead, 1.9 million pounds of cadmium and 400,000 pounds of mercury to the environment.[155] It will be recalled that exposure to these toxins is linked to severe health problems, including brain, organ, and genetic damage. This doesn't even consider the millions of pounds of plastic involved. Nor have we begun to consider the far greater number of televisions, stereos, and other consumer items containing dangerous toxic metals. In fact, our municipal landfills are as dangerous as industrial waste sites, leaching toxins into our land and water and presenting a mounting public health threat.[156]

It is now clear that this waste is most concentrated around communities without the resources to resist. Landfills, incinerators, garbage dumps, and other heavy polluters are disproportionately located near African-American, Hispanic, and Native American communities. As early as 1983, the General Accounting Office of the federal government reported that three out of four hazardous waste landfills studied were in areas with majority black populations.[157] A more extensive investigation four years later found race to be the most important variable in determining the location of hazardous waste facilities: 58 per cent of African Americans and 53 per cent of Hispanics lived in areas where hazardous waste dumping was uncontrolled.[158] The evidence is overwhelming: marginalized minority and poor communities are far more likely to be exposed to the risk of toxic waste and pollution.[159] Equally troublesome, the penalties imposed on corporations by courts for violating environmental laws were far smaller in these neighborhoods than they were in rich white neighborhoods.[160]

Hundreds of thousands of our most vulnerable citizens have paid the dearest price for our pollution. Urban poor children have been exposed to lead for decades, leading to hyperactivity, hypertension, and brain damage, and leaving them unable to perform in schools that are already overcrowded and underfunded.[161] The widespread practice of stashing toxic waste on Native American reservations has led to high cancer rates, reproductive health problems, and childhood diseases. The stretch of refineries between Baton Rouge and New Orleans is a disturbing example. Dubbed 'Cancer Alley,' these industries produce over 3 million pounds of toxic waste each year. People in the surrounding low-wage communities routinely suffer from nausea, vomiting, and skin and eye irritation; cancers rates are high.

This pattern of injustice has an equally disturbing international dimension. While wealthy countries consume overwhelming proportion of the world's products, and produce nearly 90 per cent of the world's toxic waste, the poor must deal with a disproportionate share of the consequences. Banned pesticides, hazardous waste, and toxic materials from the US and Europe, where regulations and laws are more stringent, are shipped to countries where weaker regulations and enforcement offer a freer hand to global polluters. Desperate for funds to develop their economies, feed their populations, or pay debts to wealthy governments and banks, poor countries face tremendous economic and political pressure to accept dangerous waste for little compensation. According to one estimate, 2 million tonnes of toxic debris leaves the US every year.[162] A recent two-year survey of Pacific Island countries uncovered stockpiles of hazardous waste and more than fifty contaminated sites scattered throughout the region.[163] The UN estimates that half a million tonnes of unusable pesticides are stored in the world's impoverished countries.[164] In later chapters, we will look at recent efforts to stop this practice. A recent international ban gives us room for optimism. But industry efforts to get around these new regulations also give us reason for concern. Meanwhile, in desperation, children, mothers, and workers in impoverished communities around the world drink contaminated water, breathe deadly fumes, and inhale poisonous dust created by our consumption.

More important than the actual transfer of waste to poor countries is the pervasive shift in the location of heavily polluting industries. We will look at this in more detail later. We will see how corporations that have moved to impoverished countries exploit the weaker environmental regulations and less effective enforcement. In Peru, respiratory and other health problems emerged after American corporations started to pump 2,000 tonnes of sulfur dioxide into the air each day – fifteen times the legal limit in the US. Along Mexico's border with the US, so-called *maquila* industries make products for American consumers and dump their waste into the source of drinking water of the surrounding communities. Impoverished countries do not have the resources to deal with these toxins. When 3,000 tonnes of Taiwanese toxic waste were dumped in a field in Cambodia, it was scavenged by poor villagers. Many became seriously ill; one quickly died. Local people panicked as thousands fled the city. Incredibly, Human Rights Watch reports, there was 'no law that defined, restricted or prohibited hazardous waste imports'.[165] Effective environmental regulation, exposure to deadly toxins, and the dumping of poisonous waste mirror the power and wealth inequities of the global society – the poor struggle to survive in the toxic sludge of the rich.

Once again, of course, shifting our waste to faraway places isn't a solution, even if we do not care about the health and welfare of the people who live there. It doesn't matter where the waste is; we are all affected. No one escapes. Over 200 industrial chemicals and pesticides were found in the body tissue of 95 per cent of Americans tested.[166] Every region of the world, every town, every village, is threatened. As the global environment deteriorates under the burden of our carelessness, greed and ignorance, we are all careening toward a global catastrophe.

It's going to get (maybe a lot) worse

As the world's human population rises, the demands on the planet's natural resources increase. The World Wide Fund for Nature, along with the United Nations Environmental Program's Conservation Monitoring Center, has calculated changes in humanity's total demand on natural resources – the human ecological 'footprint' – for the past three decades. They have discovered that human demands on natural resources have exceeded the earth's sustainable productive capacity since the late 1970s. In fact, the human consumption of the earth's resources now exceeds the amount that the planet can sustainably provide by a full 20 per cent.[167] In short, we are rapidly depleting the earth's natural capital. With individual consumption increasing and 86 million people added to the world's population each year, this ecological deficit will grow worse fast. The human population

1.16 / Total percentage of the earth's renewable resources used (when consumption exceeds the planet's sustainable capacity the earth's natural resources are depleted)

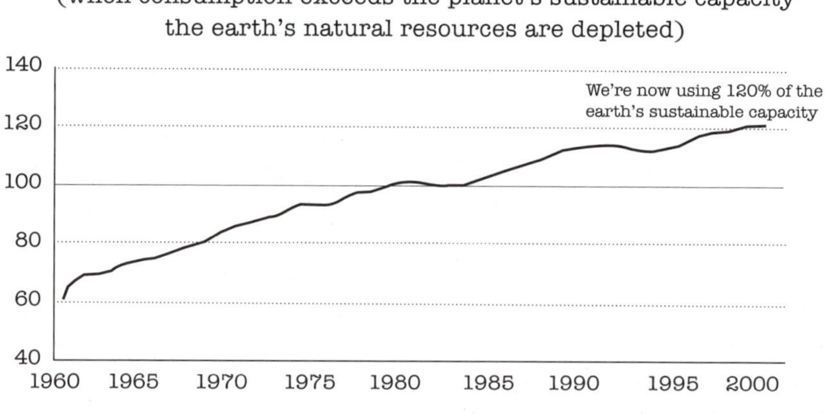

Source: Jonathan Loh, *Living Planet Report 2002*, WWF, Washington DC, 2002, p. 2.

1.17 / Global population growth (UN estimates, billion people)

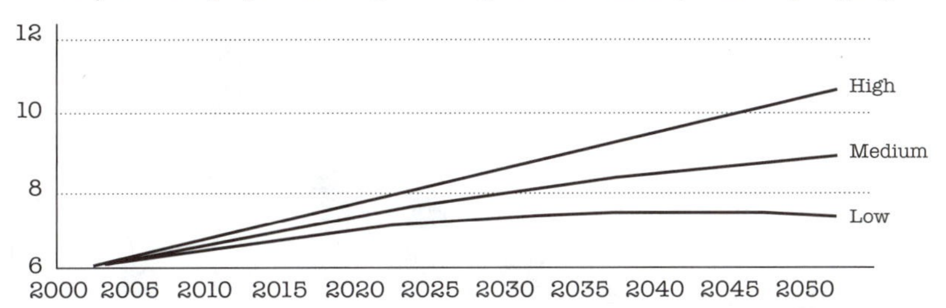

Source: *World Population to 2300*, UN Population Division, New York, 2004, p. 5, available at www.un.org/esa/population/unpop.htm.

will reach nearly 10 billion by 2050. Just to feed this population adequately, we will need to increase food production threefold.[168]

However, population growth itself is not the problem; it is only a manifestation of a much greater problem – the fundamental inequity of the world order. Rapid population growth is taking place in poor regions of the world, where an additional child provides extra income to help support the family and increases the chance that enough children will survive to take care of the parents as they grow old. In fact, the best means of reducing population is increased economic and educational opportunities, particularly for women. So, population growth can be seen as a result of the gross inequities in the world.

Yet the trouble is not really population in itself. The issue is not really about how many people are on the planet; it's about how much pressure these people put on the earth's resources. That is where things get complicated. The average American, over her or his lifetime, will account for 13 times the environmental damage of the average Brazilian and 280 times more than the average Haitian.[169] The amount of space needed to produce the food, material, and energy consumed by the average American is twelve times as great as that for the average person living in the world's poorest countries.[170] Overall, a child born in the United States will add more pollution and waste to the earth over her or his lifetime than forty to seventy children born in the impoverished world.[171] And yet it is the poor who will feel the most devastating consequences of this pollution. This is not simply a 'cruel twist of fate,' a necessary evil, or even an unfortunate side effect of a fundamentally fair system; it is built into the game – an integral component of the global system. The wealth and massive consumption of a small minority of the world's people is dependent on the poverty, marginalization, and environmental injury of a

1.18 / Unequal impacts (total ecological footprint per person, hectares)

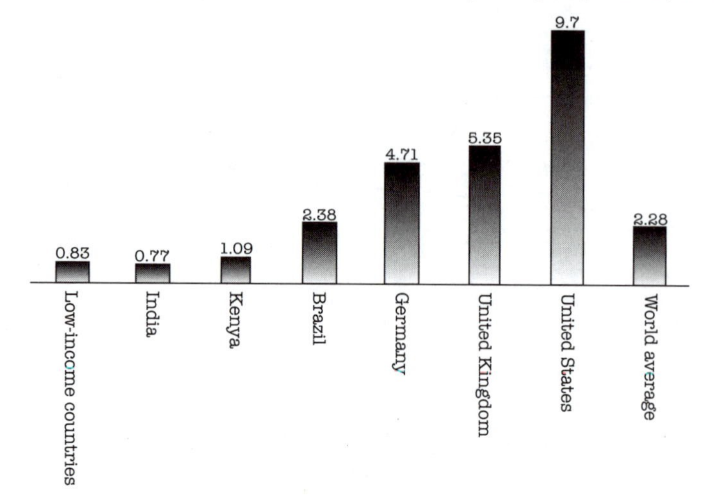

Jonathan Loh, *Living Planet Report 2002*, World Wildlife Fund, Washington DC, 2002, p. 2.

majority of the earth's population; and the system is designed to ensure that this inequity continues.

In terms of overall consumption, rather than just population, it is fairly clear that the burden of environmental devastation does not rest with the world's impoverished. While a wealthy minority overconsume, much of the earth's population are suffering from *under*consumption – lacking access to food, shelter, energy, and the resources necessary for basic livelihood. Many of the world's poor countries, trapped in a spiral of poverty, are actually reducing their consumption; in seventy countries, consumption is lower today than it was in the mid-1970s.[172] Since most countries around the world have not achieved the minimum rate of growth necessary to begin reducing poverty, suggesting that the world's most impoverished countries restrain their consumption to save the planet can only encourage greater human suffering.[173]

This does not mean that we need not be concerned with global population growth or increasing consumption and waste outside the world's wealthy countries. As the rest of the world follows the consumption patterns of those in wealthy countries, the strain on the planet's resources intensifies greatly. A global popular culture creates a voracious appetite among global elites for videos and DVDs, music, cars, clothing, fast food, and a huge variety of disposable plastic and paper waste. As the world's elite increasingly sees what were once luxury items as necessities, the environmental cost of feeding these desires will skyrocket. The percentage of the population living in urban areas is increasing in every continent, and with urbanization comes increased consumption and pollution.[174] Growth in consumption per person in East Asia, for example, is nearly three times the rate of growth in the United States.[175] Although the world's impoverished countries are far from catching up to the consumption levels of wealthy countries, with populations and consumption increasing the region seems poised to copy the mistakes of European and American histories – destroying the environment and inflicting tremendous hardship on populations for the sake of industrial growth.

China offers a disturbing example. Rapid and unregulated industrial expansion has poisoned the atmosphere, harmed the lives of millions, and intensified the overall damage to the earth's environment. Seven million people in China drink water that does not meet the country's minimum safety standards.[176] Pesticides, fertilizers, toxic emissions, and other waste from a narrow fixation on economic expansion and industrial growth have poisoned water supplies, soil, and the atmosphere. With energy consumption in this region likely to triple over the next three decades, the result is likely to be severe regional environmental damage, a powerful addition to global environmental degradation, and incalculable human suffering.[177]

More generally, as the populations of poorer countries increase and they follow our lead, their increase in energy use and waste will skyrocket. Over the next fifty years, energy consumption in Africa and Asia is expected to increase by at least 500 per cent.[178] Even a modest growth in consumption per person could have dramatic consequences. If each person in Asia, for example, were to acquire only items that many in the US would consider essential – refrigerators, air conditioners, and automobiles – the result would be an increase in energy demands that would exceed that of the United States and Europe combined. Continued population growth can only add to this stress.

The combined effect of increased population and growing individual consumption is likely to outstrip quickly not just nature's capacity to absorb our waste, but the availability of life-sustaining resources. For example, the world's oceans are being increasingly overexploited. The global fish catch has grown to over 85 million tonnes each year; the estimated maximum sustainable yield is perhaps 100 million tonnes. Our ever-increasing demand for seafood, sushi, exotic fish for our aquariums, coral and many other vital ocean resources have devastated many regions of the world's oceans and brought us closer to the sustainable limit each year.[179] Huge fishing fleets utilizing fishing lines that are up to 80 miles long with thousands of hooks that indiscriminately catch anything from swordfish to endangered sea turtles, or drift nets that are up to 40 miles long and kill everything in their path, are leaving increasingly large swathes of the ocean almost completely bereft of life. As a result, eleven of the world's fifteen major fishing areas and well over two-thirds of the world's major fish species are in decline. Pollution and the effects of ozone depletion add to the effect.

Seafood isn't the only resource that may grow scarcer. Competition over arable land could also intensify. Since the end of World War II, an area of over 3 billion acres – more than 10 per cent of the world's farmable land – has lost all or most of its productivity. Poor agricultural practices and a growing population of the desperately impoverished have radically accelerated natural soil erosion and degradation. The situation in Africa, for example, is dire. Desperate to survive, Africans overuse their soil and water, stripping the land of its natural wealth. What is left behind is often called 'desert pavement,' a rock-hard substance that is agriculturally worthless and unable to recover. Every year, 15 million acres become unproductive. Most of the impoverished world's fertile land has been cultivated, leaving rocky terrain, steep slopes, silted soil, and other marginal agricultural land. Continued overfarming will lead to increased land degradation, which will in turn intensify the crisis. The soil can take a few hundred to a thousand years to recover, if at all. The effect will not only be further deforestation to try to replace the lost farmlands – a practice that we know will be only a temporary solution – but also greater competition for the increasingly scarce land that remains.

The pressure on our dwindling uncultivated land is palpable. Africa's wild lands are projected to decrease from 70 per cent to about 55 per cent of overall land by 2015, and to 45 per cent by 2050.[180] West Asia's wild area is expected to decrease by 17 per cent by 2015.[181] Around the world, pressures on remaining wild areas will dramatically increase as climate change and the demands of an increasing population take their toll. Overall, the world's wild lands are expected to decrease some 14 per cent by 2050. While this may not seem catastrophic, it is the species-rich ecosystems that are at particularly serious risk. What will be left is mountainous, arid desert, and subpolar regions that are ill-suited for development and maintain less rich biological diversity.

Growing scarcity of resources does not end there. In many regions of the world, the supply of freshwater is growing increasingly limited. Currently some 166 million people suffer severe water shortages. Within two decades, this number will be approaching 1 billion.[182] Already, over 1 billion people have no access to safe drinking water. More than twice that number lack adequate sanitation facilities.[183] Aquifers are drying around the world and rivers are shrinking. According to one of the most ambitious worldwide assessments to date, by 2025 at least 3.5 billion people will live in areas that do not have enough water to meet their needs.[184] Even in the US, the reduction of the Ogallala aquifer under the Great Plains may cause hundreds of thousands of Americans to abandon their homes and farms in the future. But the major problem will not be in the US: China, for example, with 20 per cent of the world's population, has only 8 per cent of the world's freshwater. In India, where acute shortages of drinking water exist, 70 per cent of water sources are polluted, mostly by industrial waste.

The huge amount of water used by expanding industries is likely to intensify these shortages sharply. The production of paper, for example, uses up to 190,000 gallons of water per tonne of paper produced. Making a tonne of cotton requires 72,000 gallons. Even more disconcerting, making a single six-inch silicon wafer requires 2,275 gallons of water; so, a normal-sized computer manufacturing plant requires 236,600,000 gallons of water each year.[185] Efforts to expand high-technology industries in India have dire potential for the country's water crisis. Moreover, global climate change is intensifying water shortages. In Bolivia, for example, mountain glaciers that have acted as a natural dam, containing rain and snowfall, are receding by 10 yards a year.[186] The people of Bolivia and Peru who depend on these glaciers and the rain and snow that fall on these mountains for drinking water are likely to face serious shortages in the future. These are not the only examples; the UN estimates that over half the world's population will face water shortages by 2032.[187]

As resources become scarcer, increased violent conflict is likely. Billions of people depend on rivers, lakes, and aquifers that run across national borders. As these common water resources shrink, conflict may result. In 1975, Syria and Iraq nearly went to war over the waters of the Euphrates. Fifteen thousand died in an ongoing conflict in Punjab, India largely over river waters.[188] Israel has declared that a Lebanese plan to divert water from a river feeding an Israeli reservoir would constitute 'grounds for war.'[189] Droughts, food shortages, and flooding can not only bring strife and suffering; they can cause social instability, destabilize governments, and thus lessen our capacity to address the humanitarian and environmental crises themselves. In 1995, in a much-quoted prediction, Ismail Serageldin of the World Bank declared that 'Many of the wars this century were about oil, but those of the next century will be over water.'[190]

Conflicts are not mutually exclusive, of course; like competition over access to water, violent conflicts over oil resources are also likely to intensify. Access to oil is tightly identified with national interest, and has moved to center stage in international military affairs. In fact, the use of military power is increasingly and openly seen as a normal extension of the pursuit of resources to fuel industrial production. The United States alone spent about $400 billion on its military in 2003; that's about $14,000 for every man, woman, and child in the United States. Much of this is used to ensure continued access to oil and to control waters and lands necessary for the transportation of oil. Yet that is only part of the cost: the US spends several billion dollars more in indirect efforts to support the oil addiction of American industries. In the oil-rich nation of Colombia, for example, the US has already provided nearly $2 billion in largely military aid; recently the White House requested an additional $100 million to train and equip Colombian troops to guard an oil pipeline operated by Occidental Petroleum. The price tag for the US invasion of Iraq will certainly exceed $150 billion. This is only a small part of the billions of dollars that have been spent over the years removing uncooperative leaders, assisting often-questionable regimes, and providing weapons and training to foreign powers to ensure our continued access to oil.

The US is not the only country concerned with access to oil. Tensions in the Middle East are, of course, the most evident example. The Iraqi invasion of Kuwait in 1990 was preceded by claims that Kuwaitis were extracting too much oil from the Rumaila field. Saudi Arabia has conflicted with its neighbors over control of oil resources. The Chinese government is expanding its reach into the South China Sea, a region with large potential oil reserves. Other Asian nations are responding to ensure their own access to this oil. Russia and the Caspian region, Colombia, Nigeria, and other countries in Africa and Latin America have suffered internal violence and social strife fueled in part by competition over access to oil.

Greater conflicts are likely in the future. As consumption continues to grow out of control, oil will become increasingly valuable. Conflicting claims to these resources, intensified competition, and mounting dependency will ensure an increasingly unstable, dangerous and costly world security stage. The human cost of the ensuing 'resource wars' is liable to be great.[191] We can also expect the environmental impact wreaked by an increasingly desperate global search for oil to be absolutely devastating. The Bush administration's collusion with industry to wreak havoc on Arctic landscapes of the Arctic National Wildlife Refuge for a few thousand barrels of oil will only be the beginning. No region of the world, no pristine forest, no secluded body of water, will be safe from the efforts to wring every last drop of oil from the earth.

Yet this may not be the worst-case scenario. A recent Pentagon report addressed the potential consequences of 'an abrupt climate change scenario.' The report predicted growing global water, food, and energy scarcity. We can expect increased conflict, rioting, and even nuclear proliferation as a result, the report concludes.[192] The environmental Armageddon described in the report would dwarf the losses of 11 September 2001.

Sitting in a wealthy, protected part of the world, in the comfort of our living rooms, we might be tempted to say we can ride this out. Perhaps it's not as bad as it looks. Perhaps we can adapt. Certainly we may learn to tolerate the initial effects of environmental destruction. Perhaps we can even live with a drastically reduced number of species. Maybe wearing oxygen mask on certain days, new epidemics, skin irritations, respiratory diseases, and a lack of safe rivers, streams, and forests, could be made tolerable in the world's wealthiest countries. Maybe mass deaths from disease, flooding, or other ills brought on by our pollution of the world won't bother us if we ignore them. Perhaps, with a large enough military budget and a willingness to protect our consumption habits through violence, we can remain safe in a world torn by brutal warfare over scarce resources. But the truth is that, unless we are willing to undertake a deep and broad change in the way we do business, the most we can hope for is temporary comfort. Eventually, adaptation and ignorance will not be possible. If we do not change course, we are heading for the end. Business as usual is simply not an option.

There is something of a debate among environmentalists. Some argue that the environment is worth saving because of its own inherent value. Others say it is worth saving because of its value to humans and enrichment of our lives. In the short run, this distinction might matter. Should we protect a habitat because it has an inherent value that humans have no right to destroy? Or, should we allow it to be destroyed if the overall benefit to humans seems worth it? In the long run, however, this debate is meaningless. If we continue with business as usual, we will all die. First will go the wilderness, then will go humans. The struggle for the survival of species, the preservation of our oceans, and the protection of the atmosphere is in the end a struggle for our own survival. Global warming may contribute to the destruction of forests first, but the end point is our own destruction. Dying lakes, damaged forests, endangered wildlife are linked to disintegrating resources, economic hardship, illness, violent conflict, and humanitarian disasters. We are nature. Our shared fate rests on the protection of a delicate balance that we are rapidly destroying.

How did we get in this mess? / 2

How did we get in this mess? How did we come to construct a global system that enriches a few and impoverishes the majority? How did we build a world in which a large number do not have access to adequate food, water, and shelter while a small minority consume so much that they are destroying the global ecosystem? How did we create a system that wreaks havoc on nature's balance, destroys forests and lakes, pollutes prairies and streams, and exterminates plants and animals at unprecedented rates? How did we create a system that is not only morally vile, but that, if allowed to continue, will destroy the earth's capacity to support human life?

The troubles discussed in the first chapter are not just the result of decisions by a few politicians or even a few greedy corporate executives. It may make us feel good to blame this administration or that corporation – and usually they deserve the condemnation. However, the trouble we're in is not the result of a few bad policies or a few abusive companies. The trouble we're in is transnational – it incorporates every region of the world; it's expansive – it affects every aspect of our economic, political, and social relations; and it is systemic. It is systemic because it is not simply the result of bad policy or corrupt decision-making; the trouble we're in is actually a function of the formal rules and informal priorities that define the international economic, legal, and political order. The destruction of the environment and the degradation of human life are not accidental side effects of recent policy choices; they are entwined and unavoidable outcomes of the priorities, principles, and practices that define the rules of the global game.

So, the crisis that we're facing is not natural; it is political. It's the result not of biological limits, but of human decisions. This does not necessarily mean it is the result of *individual* human choice. Instead, this crisis is the outcome of deep-set and long-established practices and ideas that we have come to accept as normal. Over the past several centuries, we have defined a world in which the rules and practices that define our day-by-day lives – what we buy, where we live, how we work, and a myriad of other minor and major aspects of our everyday existence – threaten global ecological catastrophe. Still, human beings built this system, and human beings can do something about it. But first, we need to understand how we got here.

An unenlightened enlightenment

Human beings, of course, have always placed demands on the natural environment. Even early humans, to the extent that they were able, attempted to control their environment. The very construction of tools, agriculture, and shelters required some alternations of nature; and often the impact on local forests, soils, or wildlife was serious. The fall of the ancient Sumerian civilization, for example, can be tied to the overexploitation of natural resources.[1] However, until recently, the global human population was too small, and its demands on resources too limited, to create a lasting change to the global environment. So, while localized damage occurred, it usually did not threaten the capacity of the larger ecosystem to recover; nor did it weaken the global system's capacity to support human life.

This all changed by the mid-eighteenth century, when the human impact on the earth began to increase at an unprecedented rate. With industrialization, production of everything from bricks to cannons grew dramatically, as did the demand for fuel. In the first half of the nineteenth century, for example, world coal consumption multiplied eightfold. Greater consumption of resources destroyed larger tracts of forest and depleted natural resources at a rate that increasingly surpassed nature's capacity to recover. The growing production of waste polluted rivers, lakes, and the air. Within a few generations, the destruction would fundamentally and permanently alter the ecological balance.

These changes were justified by a new, 'enlightened' or modern way of viewing the world. Human beings were placed in a privileged position, apart from the natural world, above it, and rightfully empowered to exploit nature in any way that served their immediate interests. From this perspective, the environmental damage done was no tragedy; quite the reverse, it was in keeping with the natural order. It was simply using nature's bounty to better human life, as God and nature intended. Some argued that humanity had a sort of stewardship over the earth, and thus a responsibility to preserve it; but this view did not challenge the special position of humans in the natural order, and it was not the view of those who held the reins of power. For those who would define the new industrial order, it was the duty of humans to tame nature, to 'improve' the coarse natural environment. England's Francis Bacon, a founder of modern science, talked about his vision of nature as an economic good, without spirit or life, and, in his terms, to be 'hounded in her wandering ... bound into service ... made a slave' to the interests of men.[2]

John Locke, like Bacon a central figure in the European Enlightenment, argued that ownership itself was based on the removal of resources from the 'wasteful' state of nature. Idle natural resources – rivers, lakes, forests, and so forth – had no value. Removing these resources from nature and harnessing them to serve one's interests, Locke argued, justified the claim to ownership. 'Whatsoever, then, he removes out of the state that Nature hath provided and left in it, he ... makes it his property.'[3] When used for personal gain, nature's bounty became private property. So the exploitation of nature's resources for personal profit was itself 'naturalized' – made to appear as an expected and desirable outcome of the natural order.

This was not, of course, the starting point of environmental destruction for gain. In fact, most of England's forests had been destroyed for fuel before Locke was born. However, these ideas did set the terms for the development of the modern economic and political system that would lead to an unprecedented scope and pace of destruction and human misery, and be justified as natural, well reasoned and desirable.

Built into this world-view was a hierarchy of humanity that identified Europeans as superior and thus not only endowed by God and nature with the right to control and exploit the earth's resources but also endowed with the 'natural' right to control and exploit the earth's people. Africans, Native Americans, Asians, and even European peasants were cast as inferior kinds of humans – when acknowledged as human at all. Nature, this view declared, had equipped the noble and powerful with superior reason and intellect; so the peasantry and the impoverished – in Europe or around the world – were poor and downtrodden by their very nature. Such interpretations of the natural order were used to justify a system of privilege and exploitation – legitimating a violent, and soon to be global, system that enriched and empowered a small minority and forcibly enslaved or impoverished the majority of the earth's people.

This world-view justified an expansion of European power that was defined by its inhumanity. A brutal system of enslavement and genocide was ushered in by Christopher Columbus's arrival in the American continent and similar European exploits in Africa. Over the course of the century following Columbus's arrival, native populations of North America were violently exterminated. Entire peoples were wiped out in the space of a few generations. At the same time, millions were taken captive in Africa, if they weren't killed in the attempted capture, to be used as slaves in the Americas. This double genocide was the origin of the current global political system.

Within European society, the period from the fourteenth through to the seventeenth century witnessed the displacement of traditional social relations by a social order defined by a single mandate: profit. Social power moved increasingly away from the traditional nobility and clergy, who had defined their dominance in faith or custom, and into the hands of a new class of merchants and entrepreneurs, and then factory owners, whose power was defined by their wealth. The accompanying transformation in social values was gradual. Wealth had long been a driving motive of human affairs. However, increasingly, this quest for wealth as a goal in itself overwhelmed other aims. All aspects of social relations came to be defined in terms of property and profit. The value of things, be it land, a home, or food, came to be set only by what others would be willing to pay for them – they became commodities. This 'comodification' of society extended to human beings – the value of people was defined in what they could produce for the market.

By the late eighteenth century, land that had been shared by all for grazing, firewood, and shelter was now claimed as private property – a commodity whose value was defined only by its potential exploitation for profit. The process began in the sixteenth century. As an expanding textile industry made sheep more profitable, so landlords, bankers, merchants, and other elites conspired to take control of all remaining shared lands and reap the profits of the wool industry. Common pastures and open fields were closed off for the exclusive use of wealthy elites. Peasants who had lived on and worked the land for as long as anyone could remember were violently evicted. Rural poverty grew as peasants tried desperately to eke out their survival as laborers on what had come to be defined as someone else's land.

Traditional land conservation practices were abandoned in favor of high production for high profit. Fields that had grown enough grain for hundreds of pounds of bread were now used to produce a few sheep. People went hungry as sheep were fleeced to feed the new textile industry or slaughtered to feed the elite's appetite for mutton.[4] Lands that had remained productive for centuries as shared resources were quickly exhausted. Ancient meadows were plowed through by profit-seeking landowners; the land was permanently damaged, only to be abandoned when market prices dropped.[5]

For the powerful, removing the mass of peasantry from the land had the added advantage of creating a desperate population that would have no choice but to work in the expanding factories. Where disease and famine could accelerate the process of eviction, so much the better. So, in mid-nineteenth-century Ireland when blight hit potatoes, the staple crop of the peasantry, greedy landowners conspired with British government agents to hasten the misery, and accelerate the removal of unprofitable peasants. This tactic would be repeated often.[6] Human beings became complete commodities: when not profitable, they were eradicated; when needed as labor, they were essentially bought and sold on the labor market, paid just enough to keep them alive but little enough to keep them desperate. People who had lived with the cycles of nature for generations were now fodder for the machines of industry, their connection to the land lost.

The food produced on the land on which the peasants once lived was now sold to them at a profit. Industrial 'efficiency' – defined by the greatest production of profit, not the greatest good to the people – led to a rise in the production of food. However, this was coupled with a dramatic rise in its price, so that between 1780 and 1800 the price of wheat tripled.[7] Workers would work longer hours, in brutal conditions, and could barely afford enough food to survive. While the wealthy justified the transition in the name of 'progress,' these changes must hardly have looked like progress to the majority who suffered greatly.

For a privileged few, however, progress did seem apparent. Industrialized societies were now capable of producing more than the basic necessities of life. The impoverishment and exploitation of the majority allowed a privileged few to enjoy luxury goods. This rising class replaced the old nobility and defined a new system based on industry, consumption, and the endless quest for material wealth.

Industrializing nature

This transition to capitalism offered little to the vast majority of the population, who were now cast as workers rather than peasants. Of course, before the Industrial Revolution nobles had often brutally exploited peasants and survival had often been difficult. What was apparent by the nineteenth century was that the new system based on commerce and property was equally brutal and every bit as absolute. Since having many poor people would drive down the price of labor, in the eyes of the wealthy and powerful there could not be too many impoverished workers. Indeed, greater profit required dire poverty, for only with the threat of dire poverty would the marginal be willing to continue working at starvation wages. No one was spared. Children and mothers desperately sold their labor to the highest bidder. Families that had lived by keeping a few animals and a small garden while spinning wool, collecting kelp, or some other activity to be able to purchase the few goods they needed, were torn from their homes and filed into newly created and burgeoning cities to fill the labor market.

For those jobs that were too brutal or degrading for even the most desperate, slavery provided an answer. While some form of slavery had been around for a long time, industrialization fostered a mode of slavery that was more total and inhuman than any previous system. The slave trade defined an industrialized and inhumanly vicious system of exploitation that magnified all the barbarities of the industrial order and revealed the brutal results of placing profits over humanity. Africans were cast by their European enslavers as subhuman, indolent, and incapable of proper industry if left to their own devices. Thus, like nature itself, Africans had to be forced to proper utility – that is to say profit – through even the most brutal means. The expansive slave system that developed would destroy the lives of hundreds of millions of Africans and obliterate countless African cultures and societies.

Those who could not be profitably exploited, such as Native Americans, were simply eradicated. In fact, a principal claim used to justify European genocide against the native peoples of America was that they had not developed private property. The European legal scholar Emmerich De Vattel defended European conquest by arguing that the Native Americans' 'unsettled habitation in those immense regions cannot be accounted a true and legal possession; and the people of Europe … finding land of which the savages … made no actual and constant use, were lawfully entitled to take possession of it'.[8] In short, because they had not claimed the earth as private property, because they had not imposed the unsustainable agricultural practices and exploitative social structures that private property had brought in Europe, the native peoples could rightfully be destroyed. An Indian world-view that saw the earth as a gift to be shared was displaced by a European view of the earth as a source of commodities and wealth. Neither the land nor the Indians could withstand the changes this entailed. Climatic changes, devised land scarcity, soil depletion, epidemics, the severe reduction in wildlife and plant species, deforestation, forced migration, and violence took a severe toll on the native peoples and their cultures.[9] The barbarity with which this was undertaken was linked to a racist world-view that cast the 'animals, vulgarly called Indians' as base and violent species. Like the natural world, Indians were to be controlled, manipulated, or destroyed to suit the interest and pleasure of white men.[10]

Once again, when disease hastened the pace of the genocide, so much the better. Any increased efficiency in granting European conquerors access to natives' land was welcome and, their self-righteous argument went, a manifestation of God's will. As John Winthrop, leader of the New England colony in the 1630s, reported, the Indians were 'neere all dead of the small Poxe, so the Lord hathe cleared our title to what we possess.'[11]

Industrialization dramatically reorganized every aspect of society. Traditional town relationships were torn apart. Small-scale artisanship was replaced with the crude, energy-intensive but more profitable mass production. The steam engine allowed production to take place in concentrated areas, rather than being dependent on the power of rivers; so industry moved away from water sources and to inland towns and cities close to labor, supplies, and markets. These transformations were also associated with a notable increase in the rate of global population growth; and, in its most simple sense, more people meant more demands on the environment.

The shift to industrialization didn't just mean an increase in the demands on the earth's resources, but a qualitative shift in how the ecosystem was used – the atmosphere became an exhaust dump, the earth became a quarry for profit, rivers and lakes became sewers for the waste of industries. As the demand for coal increased from 27 to 230 million tons per year in the mid-nineteenth century, the devastation of coalmining spread across Britain, Europe, and the US.[12] Coal-powered industries pumped massive amounts of toxins into the air and spread heavy metals and other dangerous pollutants over mountains, lakes, and rivers far downwind.[13] The smog that hung over industrial centers was palpable. In 1880, more than 2,000 people died in just three weeks due to particularly severe

2.1 / A sharp growth in population (millions)

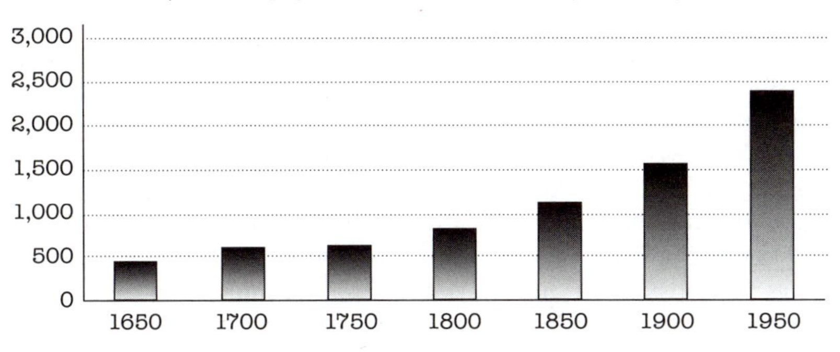

Source: 'Historical Estimates of World Population', www.census.gov.

smog.[14] Factories were built along rivers to allow for the easy disposal of massive waste, creating vast stretches of poisoned water. Lakes and rain in Britain grew more acidic.[15] The replacement of natural fertilizers such as manure with artificial fertilizers contaminated the countryside and waterways. The toxic contamination of the soil and water left by these industries is still evident in our rivers and countryside.[16]

In urban areas the environmental costs were particularly clear.[17] People lived in the smallest spaces, without drainage or sewerage, without clean water. Filth and diseases like typhoid and cholera were rampant. The massive burning of coal in foundries, factories, and homes poisoned the atmosphere and engulfed cities. The stench from polluted rivers and streams was often unbearable. The expansion of industrial diseases ranging from mercury or lead poisoning to respiratory diseases was rampant. In some industries cancer killed one in four workers. It is no surprise that the death rate in cities far outpaced the rate in rural areas.[18] In the industrial city of Manchester in 1840, 60 per cent of working-class children died before their fifth birthday.[19]

Friedrich Engels provided a detailed account of the public health catastrophe caused by severe poverty and pollution. His description of a working-class district in Manchester offers a chilling picture:

> The method of construction is as crowded and disorderly.... Right and left a multitude of covered passages lead from the main street into numerous courts, and he who turns in thither gets into a filth and disgusting grime, the equal of which is not to be found – especially in the courts which lead down to the Irk [River], and which contain unqualifiedly the most horrible dwellings which I have yet beheld. In one of these courts there stands directly at the entrance, at the end of the covered passage, a privy without

> a door, so dirty that the inhabitants can pass into and out of the court only by passing through foul pools of stagnant urine and excrement.... Below it on the river there are several tanneries which fill the whole neighbourhood with the stench of animal putrefaction. Below Ducie Bridge the only entrance to most of the houses is by means of narrow, dirty stairs and over heaps of refuse and filth ... The view from this bridge ... is characteristic for the whole district. At the bottom flows, or rather stagnates, the Irk, a narrow, coal-black, foul-smelling stream, full of debris and refuse.... In dry weather, a long string of the most disgusting, blackish-green, slime pools are left standing on this bank, from the depths of which bubbles of miasmatic gas constantly arise and give forth a stench unendurable even on the bridge forty or fifty feet above the surface of the stream.[20]

Nevertheless, political leaders saw the national interest, and thus their task, tied to the enrichment of industry and trade, not the protection of the environment or the population. The distinction between industrial leaders and state officials blurred as the fixations with industrial profits were cast as entwined with national greatness and state power. Government officials, nominally accountable to the people, worked to ensure the continued profit of the new wealthy class. Public money was used to build a transportation infrastructure, otherwise subsidize industrial ventures, enforce property laws, and supply a workforce that would ensure industry's profitability and security.

The North American landscape was deeply wounded by the advance of industries. The construction of the Erie Canal set off a great mania of canal digging. Forests were devastated to feed the steamboats that traveled these canals, often leaving large swaths of deforestation around canals and major rivers. When railroads replaced canals, the environmental consequences dramatically increased. Railroads left deep scars on the countryside and dramatically increased demands on forests and mines for fuel and materials. More importantly, railroads expanded industry and its environmental consequences to the far reaches of the continent. Most notably, cheap rail transportation enabled bulk, centralized production, increasing the transportation involved in the production of a wide array of food and industrial products; and so the energy requirements of manufacturing grew, increasing the environmental consequences of an array of daily items.[21]

The costs paid by wildlife for American industry was evident very early. The terrible effects of deforestation were apparent by the early 1800s and increased dramatically over the following century.[22] In 1880, the Chicago lumber market was consuming over a billion foot of board every year.[23] Laws intended to secure the land claims of individual farmers in the West were used by corrupt and manipulative timber corporations to take ownership of huge areas of forest, remove the most valuable trees without replanting, and leave behind a distressed landscape. The enormous and thousand-year-old redwood forests of California's coastal ranges were devastated. The Great Plains – majestic but unprofitable – were surrendered to cattle grazing, destroying the natural ecosystem and leaving sagebrush and degraded soil.

Hunting combined with this destruction of natural habitats to overwhelm wildlife. An already stressed population of perhaps 40 million American bison was decimated within a few years through habitat destruction, disease, and hunting. Killed for their meat originally, Bison were then slaughtered more aggressively for their hides, and often just for sport.[24] A favorite target of hunters, the grizzly bear was extinct in the California Sierra Mountains by the 1920s, a quarter of a century before it was made the official state animal! One of the most notable cases of mass annihilation was the passenger pigeon. In early America, huge flocks of passenger pigeons filled the skies; and the population certainly numbered in the billions. By 1914, thanks to mass slaughter for sport, decorative feathers, and meat, the last of these blue-backed doves died in captivity. As early as 1864, the early ecologist George March Perkins noted that wherever humans set foot 'the harmonies of nature are turned to discords'.[25]

For corporate elites, ecological concerns simply did not exist – they had no standing in the marketplace, and thus had no meaning for the entrepreneur whose sole drive was maximizing profit. Nature's bounty was only valuable to the degree it was made to serve human profit; so no concern was given to the environmental damage wrought by the quest for wealth. The dramatic destruction caused by gold and silver mining in the West offers one of many possible examples. Mining industries fixated on the potential for great profit, and, given a free hand by laws that allowed unrestricted access to any who could stake a claim and make a profit, used high-pressure water to tear through mountainsides, sweeping away thousands of tonnes of soil and rock each month, destroying wildlife, washing away topsoil, and leaving only debris and barren hillsides. When this hydraulic mining was replaced with hard-rock mining in the late nineteenth century, the destruction became worse and more long-term.[26] Rock was pulverized, and mercury was added to bond with the gold. The result was not only the destruction of majestic landscapes but also the contamination of rivers, lakes, and soils with toxic mercury. That mercury still sits in many rivers and streams and presents a hazard to this day.

A global system

New forms of industry fostered new forms of global exploitation. Europeans had once looked around the world for precious metals, spices, and slaves. With industrialization, demands for raw materials drove an even more complete form of global exploitation. Slavery was not so much displaced as transformed into an overwhelming system of control that made entire peoples and their resources slaves to the industries of the wealthy. Africa, Latin American, and Asian countries were incorporated into a global system as the source of raw materials to feed European and American industries and as a market for their manufactured goods.

Nothing made the close association between industrial interest and military force more clear than the colonization of India. The very agent in charge of the colony, the British East India Company, was a profit-seeking enterprise – a monopoly venture supported by the British state and industrial elites. The entire economy and industry of India were geared to meet the needs of British industries. India was flooded with British goods, destroying local enterprises. Traditional cotton weavers, for example, were overwhelmed by cheap imports from British industry. Traditional practices that had functioned to sustain natural resources and social welfare were displaced by a British system focused on the extraction of maximum profit. For example, prior to British conquest, peasants paid 30 per cent of their harvest to a communal fund used to maintain common village roads, irrigation, and other resources. After British conquest, payments more than doubled, and the money was taken as profit by the East India Company, not reinvested into the local community. The result was a peasantry enslaved to the demands of European industries and the destruction of the communal commons that had allowed for centuries of sustainable farming. Whatever stake local communities had in the preservation of local water sources and soil quality were weakened when the colonizers became sole beneficiaries of these resources.

This brutal global system of exploitation was justified in the most perverse ways. For example, when the Chinese emperor blocked the sale of opium that was being circulated throughout China by the British East India Company, the British military intervened in the name of 'free trade.'[27] The British company had been searching for a way to gain access to Chinese resources. Trade in highly addictive opium offered them the possibility to create an addicted and intimidated population that, once dependent, would sell their products cheaply. British entrepreneurs hoped this would allow them to exploit fully the resources of China without the tremendous cost of mounting a military campaign. When the Chinese emperor objected, the British seized several Chinese cities. Later concessions forced the emperor not only to allow the British a free hand in China but Chinese emigration to serve as cheap labor in the US and South Africa. The lesson was clear: the entire world, even those parts that had not been directly colonized, was subject to the European demand for profit and power.

Indeed, isolation from this expanding system of control was not an option. Somewhat after the Opium Wars, when Japan attempted to remain inaccessible to American industries and traders, this point was clear. American ships sailed into Tokyo Bay to force Japan to open its markets to the West, surrender control of Japanese trade, and grant Westerners immunity from Japanese laws. When social unrest ensued, US and European ships attacked important Japanese fortresses and helped install a Western-style government friendly to US businesses.

The great transformation to industrial society that had occurred in Europe with such devastating environmental and human consequences was now repeated in even more brutal terms around the world.[28] Croplands that had produced small-scale agriculture for local consumption were now consolidated under the control of wealthy landlords, who, echoing previous events in Europe, removed humble farmers from the land and produced crops for the much more profitable export market. Fewer and fewer wealthy elites controlled more and more land. The result was a global population of impoverished peasants without the means to provide for their families' survival.[29]

The full barbarity of this global order was most infamously evident in the Belgian Congo. Anxious not to miss the divvying up of Africa by European powers, King Leopold II of Belgium created his own personal colony in Africa as a private entrepreneur. With extravagant tastes to serve, Leopold II ran a brutal system of forced labor and amassed a fortune.[30] Village women and children were rounded up and held hostage, forcing the men of the village to gather natural rubber as ransom for their families. Entire villages were slaughtered by colonizers to ensure a properly terrorized population and thus the continuation of profits. Individuals who failed to meet their work quotas had their hands cut off. While Leopold's example was particularly heinous, it was not alone. By the last decades of the nineteenth century, nearly all of Africa had been claimed by a European power, using violence and intimidation to transform brutally the continent's economy and society to suit Western interests and feed Western industry.

Resistance was violently put down. The German repression of the Herero and Nama opposition in Southwest Africa was one of many examples.

Germany had colonized the region with the intent of raising cattle on the land for the European market. Africans were forcibly removed from land and relegated to work camps to make way for European settlers and cattle. When Africans rose in resistance, they were viciously put down by the German military. Resisters, suspected resisters, or potential troublemakers were killed, thousands were taken to death camps, entire communities were annihilated; the campaign of terror ended with roughly 80 per cent of the 80,000 Herero exterminated. The survivors were relegated to landless poverty, forced labor, and inhuman living conditions.[31]

With virtually every region of the world under the direct or indirect control of Western powers, the Western economic system became truly global. Europeans enjoyed products from around the world. Wealthy Europeans and Americans wore cottons and silks from Africa and Asia, drank Chinese tea or South American coffee, furnished their homes with exotic woods from Burma or Malaya. These elites enjoyed the benefits of having the economies, societies, and governance of Asia, Africa, and Latin America violently transformed to suit the demands of Western industry. Crops, mining, and manufacturing around the world were organized to produce for markets in Europe and the US. In addition to the sheer brutality through which this system was imposed, the resulting global economy left the rest of the world tremendously exposed to changing demands or preferences of wealthy Europeans and Americans; and an ever more severe exploitation of natural resources was required to continue to feed the growing consumption of the European and American markets.

The environmental consequences were dramatic. African populations that had lived in balance with their environment were radically altered by the imposition of colonialism.[32] Colonial powers shifted local harvests toward the production of crops for European consumption and European profit. Local food shortages resulted. While profitable for European colonists, these crops were often not appropriate for the local ecosystem; water-intensive crops were planted in dry regions, monoculture displaced traditional crop rotations. The repeated production of a single crop for export, rather than the traditionally rotated local crops, exhausted the soil and increased susceptibility to pests. The result was often land degradation and damage to the local ecological balance.[33]

As early as the mid-eighteenth century, forests in India, Burma, Fiji, and other Asian nations were seriously degraded or destroyed.[34] When the United States gained control of the Philippines, modern industrial logging was introduced to the region. In less than a century, more than two-thirds of the original forests were destroyed. European mining ventures in Asia and Africa destroyed vast tracts of the natural landscape and led to food and water shortages. The expansion of plantations for export crops in Latin America had similar outcomes. American industrialists undertook the 'Conquest of the Tropics,' asserting their intent to bring civilization to savage peoples and productivity to wasted nature. With the cooptation of local elites, they cleared forests, displaced indigenous tribes, condemned local farmers to landless peasant status, and constructed a massive empire of commerce to produce sugar, bananas, coffee, rubber, beef and lumber for the world's wealthy.[35]

In short, enrichment of European and US industries was accomplished through the violent social and ecological impoverishment of much of the rest of the world. With global industry concentrated in wealthy countries, Africa, Asia, and Latin America were turned into sources of cheap food and raw material to feed these profit-making machines. As a result, over the course of the nineteenth century, Europe's and the United States' share of the world's manufacturing output rose from 30 to 89 per cent.[36] The claim of free trade was clearly hypocritical. Imperialism established a monopoly on the terms of trade and was intended to close off global markets to competitors. Tariffs and other measures protected domestic industries in America and Europe from possible competition, while the major industrial powers forced the rest of the world to open its markets. What emerged was not a level playing field for free trade but a strongly organized and violently enforced system of controlled trade that benefited industries in Europe and the US and actively impoverished peoples in Asia, Africa, and Latin America.

US dominance

By the end of the nineteenth century, technological, social, and political changes had made the global system more expansive and complete. Oil, electricity, and steel emerged as dominant commodities. Enterprises grew more extensive. International marketing and distribution networks expanded. All of which meant a greater demand for nature's resources. Industries had become increasingly dependent on oil not only to lubricate their machinery, but also to drive their power plants and transport their products. As military hardware modernized and modes of warfare grew to encompass the whole of the nation's industry and population, the great powers – Britain, France, Germany, Russia, and the United States – began to see access to oil as crucial to their political and economic security.

The rush for oil wreaked environmental destruction in oil-boom towns across America, presaging the global havoc that would arise a few decades

2.2 / Carbon dioxide emissions, 1800–1920 (million tonnes)

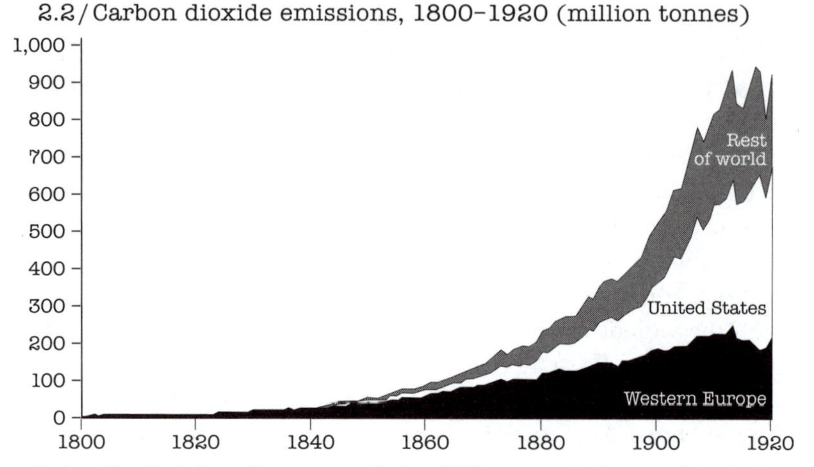

Source: Carbon Dioxide Information Analysis Center, US Department of Energy, http://cdiac.esd.ornl.gov.

later. Common law gave ownership of oil to the drillers, who had a strong incentive to take as much as they could as quickly as they could.[37] The toll on forests, rivers, lakes and wildlife was dramatic. Oilfields expanded out of control, tearing through forests and choking rivers with their waste. The full consequences would not be recognized for several decades. The sharp rise in carbon emissions at this time would set the early course for the current global climate change emergency.

American industries grew, fed by the tremendous natural resources of the continent and driven by an expanding mass market. Large chain stores grew at unprecedented rates. In major cities, more than half of all retail products passed through chain stores by the end of the 1920s.[38] Sears, Roebuck and Co., Montgomery Ward, A&P, and others began to expand their share of the market as small stores felt the pinch. Middle- and upper-class customers turned to these stores for cheap amenities and convenience items, phonographs, fashionable clothes and the like. Increasingly, the conveniences that had been available only to the wealthy were more commonly attainable. Household appliances and cars were now the norm for a small but growing middle class. The working classes followed suit as best they could afford, enjoying having more than the bare necessities of life. Thus the loyalty and support of a relatively small middle class, and the general acquiescence of a larger working class, were bought with cars, Sears catalogues, and other charms of the industrial age. Of course, this was only possible because of the terrible exploitation of peoples and nature around the world. The global triumph of the system allowed for the cooptation of those who were most able to change it.

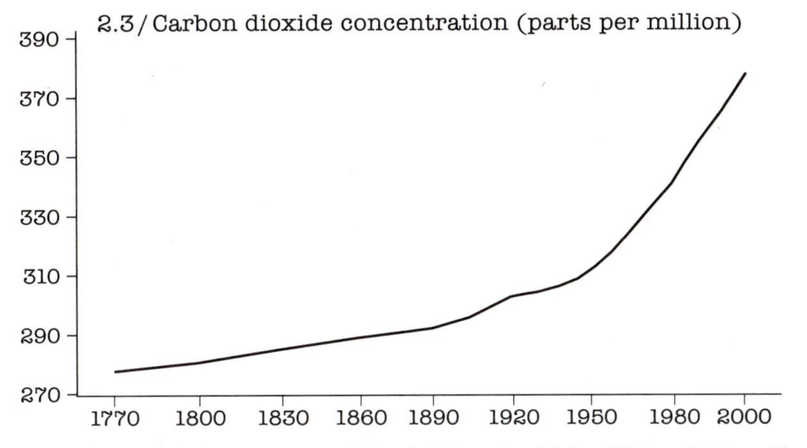

2.3 / Carbon dioxide concentration (parts per million)

Source: A.M. Mannion, *Global Environmental Change: A Natural and Cultural History*, Longman, New York, 1997; Clive Ponting, *A Green History of the World: The Environment and the Collapse of Great Civilizations*, Penguin, New York, 1993, p. 384; Carbon Dioxide Information Analysis Center, US Department of Energy, http://cdiac.esd.ornl.gov.

The food industry provided a clear example of the dramatic increase in the environmental costs of mass manufacturing. The production, distribution, and sale of beef, pork, and grain were amalgamated, industrialized, and fed by fossil fuel. This meant the destruction of forests and wetlands for intensive cultivation. It also meant consumption of more oil in the production of synthetic fertilizer, as well as the pollution of land and waterways by these fertilizers and the other waste of these massive industries. In fact, the packing and shipping of pork and beef were infamous for their filth. Rivers grew putrid with the body parts and fluids from animal carcasses; burning waste at the plant poisoned the atmosphere and produced an intense stench.

The cost of expansion without concern for ecological balance was clear in the early 1930s when one of the periodic droughts that had occurred historically in the Great Plains brought devastating consequences. Tens of millions of acres of the dense grass cover that had kept the soil of the Great Plains stable for centuries had been plowed up. Without its protective cover, the exposed, fragile, dry soil was swept up by high winds. Several hundred million tonnes of topsoil were carried away and deposited hundreds of miles downwind, leaving the 'Dust Bowl' in its wake.[39] By 1938, 850 million tonnes of topsoil were being lost every year. Massive dust storms engulfed the plains. While much of the region would eventually recover, hundreds of millions of acres of American cropland remain seriously damaged or ruined.

While the Russian Revolution of 1917 might initially have seemed to promise an alternative for the world's downtrodden, such hopes were quickly dashed by a corrupted regime that transformed a call for popular justice into a legitimating instrument for dictatorship. The result was a level of oppression and environmental catastrophe unparalleled in world history. With an absence of colonies to exploit, the new Soviet regime turned to the natural resources of the interior and the exploitation of the peasantry in a brutal program of forced industrialization. The violent collectivization of farms entailed the deaths of millions through both murder and starvation. In a desperate effort to keep up with Western industrial growth, the Soviet government brutally suppressed the population, exploited workers, and devoured nature's resources. Massive hydroelectric projects built by prison labor, huge and inefficient heavy industries, a transportation network built for political priorities, and many more such blights expanded without the least regard for ecological damage. Rivers were turned to toxic sludge, huge tracts of forests were cleared or flooded, air pollution engulfed regions and acidified surrounding waterways, all in the name of the 'Soviet people.'[40]

In the United States, the Great Depression and the mass mobilization of the working class did lead to reforms in the system. New Deal legislation regulated wages and prices, empowered unions and provided retirement benefits, job programs, and unemployment relief for American workers. These measures were coupled with laws regulating work hours, safety conditions at work, child labor, and minimum wages. The formulation of all these measures was dominated by big business to be sure; and all these reforms must be seen as efforts to stabilize the existing system for its own protection in the face of massive popular unrest. Nevertheless, the lives of American workers did perceptibly improve with these reforms.

In fact, some of the programs that were heralded as the greatest achievement of the New Deal were catastrophes for the environment. Mills produced tonnes of coal waste. Refineries dumped chemicals into rivers and harbors or allowed them to seep into groundwater. Agricultural subsidies paid farmers not to produce food and accelerated the move toward total corporate control of agriculture. Massive projects were designed to provide cheap power for the country's industries and growing population. The Tennessee Valley Authority regulated water and electricity throughout the region through the use of enormous efforts to redefine nature to our liking. Boulder Dam blocked the Colorado river. Bonneville dam blocked the Columbia river. Throughout the American West, the result was enormous ecological damage and a redistribution of power to industry elites and government bureaucrats beholden to them.[41] In an ultimately unsuccessful struggle to block the flooding of the magnificent Hetch Hetchy valley in Yosemite by such a dam, the naturalist John Muir noted the priorities of the age with disdain. 'These temple destroyers, devotees of ravaging commercialism, seem to have a perfect contempt for Nature, and, instead of lifting their eyes to the God of the mountains, lift them to the Almighty Dollar.'[42]

American affluence and global catastrophe

The end of World War II confirmed the culmination of two trends. First, the United States was clearly established as the major industrial and military power in the world. After the war, with much of Europe worried about the hostile intentions of the Soviet Union, the United States constructed an international order that would provide stability and protection to global capitalism, but always on terms that were favorable to its own interests. Second, a deep alliance of government and industry in the United States was in place. So, when the CEO of General Motors, Charles Wilson, was asked why he did not sell his GM stock when he took the job of secretary of defense, he replied with a straight face: 'What was good for the country was good for General Motors, and vice versa.'[43] President Eisenhower warned Americans of the overwhelming power of the 'military–industrial complex' in 1961; but by that time the terms were set. The nation's very fate, it was argued, was tied irrevocably to industrial growth; and so it was the job of government to ensure the continued expansion of corporate power and profits, even at the expense of human lives.

Growth became the mantra of the second half of the twentieth century. Although there was no apparent reason that continued expansion was necessary, a kind of faith developed, based perhaps in the voracious desire for increasing profits, that if production and consumption did not continuously expand, society would suffer. The choices came to be cast as economic expansion or social misery, when in fact social misery was often caused by the blind drive toward expansion. For corporate and political leaders, and much of the general population, the world had come to be seen as competing national economies. In a global competition fixated on growth and measured with a yardstick that narrowly focused on industrial profits, government leaders increasingly began to see their chief task as the encouragement and facilitation of continued consumption and profits.

The United States took the lead in designing and installing a series of global institutions that entrenched Western economic dominance and ensured terms of trade and commerce that were favorable to Western industries. The US dollar was enshrined as the dominant international

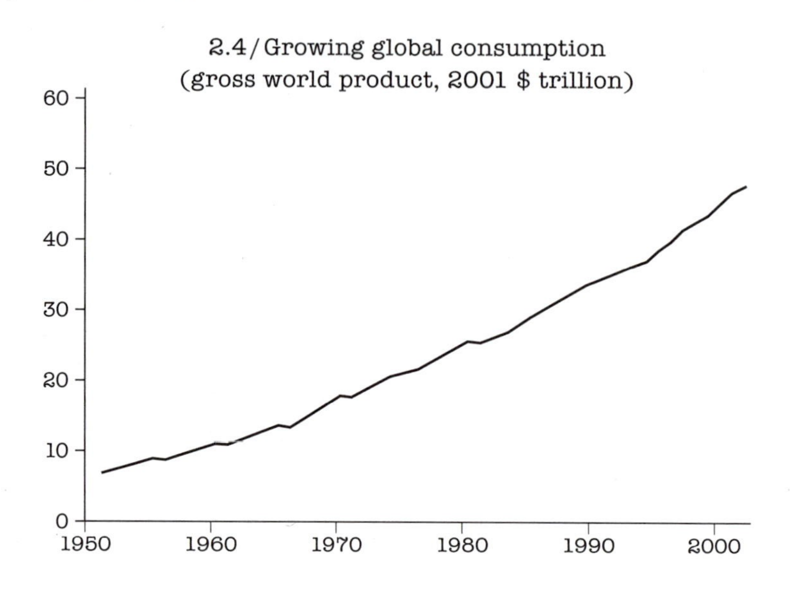

Source: Michael Renner et al., *Vital Signs 2003*, Worldwatch Institute, Washington DC, 2003, p. 45.

currency. The International Monetary Fund (IMF) was established to monitor and enforce a system of exchange rates and convertible currencies. With the US providing nearly one-third of the resources, they reasoned, they should also have the lion's share of the voting power in IMF decision-making. The IMF thus enforced the desire of conservative US financial and industrial leaders, with priorities on stability of the system and continued profit. Initial efforts to form an International Trade Organization were scrapped by the US when it became clear that the institution would favor poor countries at the expense of US interests. The General Agreement on Tariffs and Trade (GATT) emerged as the chief international trade organization with the aim not of free trade but of regulated, *freer* trade – carefully controlled and limited to protect the interests of wealthy countries. For example, agriculture, a key US industry needing protection from competitors in poor countries, was excluded from GATT negotiations so that wealthy countries could continue to protect their uncompetitive industries. In the process, of course, this seriously impeded the growth of impoverished farmers around the world. So, while explicitly in favor of free trade, these institutions in fact helped manage a global system of controlled trade, with terms defined in the favor of the wealthy.

2.5 / Growing fossil fuel addiction
(world fossil fuel consumption, million tonnes of oil equivalent)

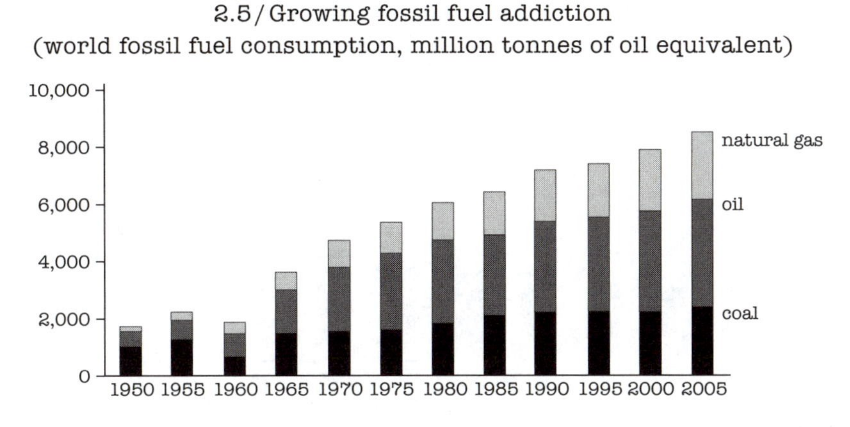

Source: Michael Renner et al., *Vital Signs 2003*, Worldwatch Institute, Washington DC, 2003, p. 34.

An alliance of corporate and government interests ensured access to the massive amounts of oil needed to fuel the American economy. Over the previous half-century, oil had displaced coal as the fuel of industrial growth. The expansion of the automobile as a central fixture of American society and industry confirmed this shift. And as early as 1946, when the US mobilized to force the Soviets out of Iran, it was clear that the US was ready to use military force to ensure its unimpeded access to cheap oil. When the US helped install the brutal Shah of Iran in 1953 to ensure its control of the region, it was apparent that continued access to oil trumped concerns for democracy or human rights.

Manufacturers who were anxious to maintain the growth and profits of the wartime boom promoted the continuation of military production and shifted to the manufacturing and promotion of cars to fill the void. The demand for petroleum grew dramatically as a fuel for cars, as a material for the construction of roads, and as fuel for the machinery that built the roads, bridges, overpasses, and parking lots of the car-centered society. New industries emerged to produce the tires, glass, steel, and host of materials and products needed to support the nationwide fixation on the automobile. By the end of the century, one-third of all cars in the world would be owned by the 4 per cent of the world's population who live in the US.[44]

Individualism and consumption, embodied in the automobile and growing suburban sprawl, defined the terms of this new America. All urban planning revolved around the automobile, with massive amounts of space and environmental resources used to ensure auto access and parking.[45]

2.6 / Convincing people they want what they don't need (advertising expenditure, $/person)

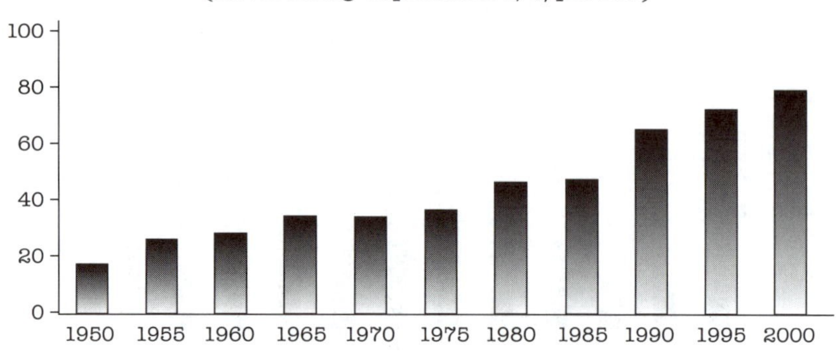

Source: Michael Renner et al., *Vital Signs 2003*, Worldwatch Institute, Washington DC, 2003, p. 48.

Life in the new suburban America was impersonal, undifferentiated, and dependent on tremendous waste in resources. The environmental space required for each family shot up. Homes grew, as did heating, cooling, lighting, and maintenance requirements. Every foray outside the home required an automobile. In the 1920s, Detroit had been the first city to decide to widen roads rather than build a subway system; by the 1950s, such decisions were a matter of course.[46]

This was not a natural development of modern society; it was a designed and manipulated project on the part of industrialists with an interest in expanding profits. In particular, General Motors, Standard Oil and Firestone aggressively bought up public transportation systems and closed them down, eliminating more environmentally sustainable competition to the automobile. Over one hundred commuter rail systems were bought and shut down. Thousands of miles of rail lines were ripped up. Intensive lobbying and influence peddling ensured that a massive highway system was built with public funds to replace them. The environmental and social costs were staggering.

The growth of consumerist culture was neither natural nor unproblematic. It had to be aggressively promoted and manipulated by corporate leaders. Politicians desperate for public approval and the financial support of industries found a solution to both needs in the aggressive promotion of economic growth and greater consumption. Marketing grew into a major industry with a single aim: convince people to want things that they do not need. Spending on advertising increased by more than a billion dollars a year.[47] Corporate interests were in sales not sustainability. It was better to produce a cheap product that could be replaced in a few years than a more durable product that might saturate the market. Hence, relatively short-term obsolescence was engineered into the product. Manipulations of style or fashion grew more important than quality or durability as manufacturers struggled to maintain continued expansion of sales.

Marketing did its job; blind consumption became an American pastime. Americans chose to work more hours so they could consume more.[48] Mass-produced consumer devices expanded dramatically. The American appetite for electrical appliances, clothing, toys, and an innumerable number of other products grew. And the energy demands needed to produce, distribute, sell, and market these items added to the growing environmental tab. Food was prepackaged, and marketed for esthetics, convenience or style rather than nutrition. Clothing was no longer a necessity but a cultural fixation requiring multiple new wardrobes each year. Perfectly adequate and inexpensive products were replaced with energy-intensive, stylized alternatives. Electric toothbrushes, hairdryers, pencil sharpeners, and a host of other conveniences replaced environmentally benign practices with energy-intensive, environmentally costly alternatives. Cheap plastics replaced more environmentally friendly and energy-efficient materials. Homes were no longer about shelter but about status and style. Consumption itself seemed to replace happiness as a national pursuit. Indeed, while the number of Americans calling themselves happy peaked in the 1950s, consumption continued to increase at a dramatic pace.[49]

Yet, to many, progress seemed apparent. Many of the citizens of wealthy countries achieved a material standard of living that had been unthinkable a few decades previously. Health care had improved and became more available. Education was more widespread. Hunger and social hardship seemed less common. The grinding poverty that had plagued large sections of the population seemed to become the plague of a minority. Yet such progress was unequally distributed; and poverty remained a persistent problem among this plenty. Differences among regions and races was great. A system of functional racial and class apartheid was sharpened as wealthy whites moved to sanitized, safe suburbs, leaving the poor and marginalized in the declining inner cities (inner cities that were now stripped of adequate public transportation facilities).

The expansion of consumption meant the expansion of waste and pollution. Industries emitted carbon dioxide, nitrogen oxide, hydrocarbons, and other particulate matter in unprecedented amounts. Gary, Indiana, a center of the steel industry, became one of the most polluted places on earth.[50] The rivers of the area became sewers for the waste of car culture. Each year, hundreds of thousands of barrels of hazardous waste

2.7 / Global energy consumption (real and projected, exajoules)

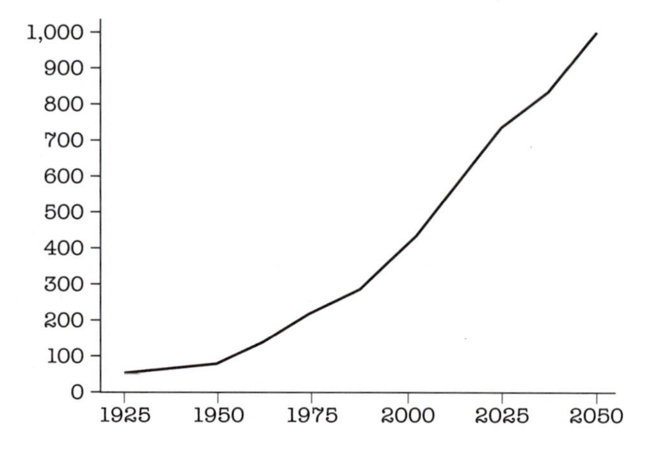

were dumped into the surrounding landscape. Toxic waste sites – located disproportionately near poor, marginalized communities – cropped up around the region.

The human costs mounted, but were often effectively hidden by industries and co-opted public officials. In October 1948, for example, a small zinc factory town in Pennsylvania was engulfed in a dense, poisonous fog for over a week. Within a day, eighteen people had died. More than 6,000 fell sick.[51] Concerned that public awareness might force them to clean up their act, industry leaders orchestrated a whitewashed government report that laid blame principally on the weather.[52] A few years later, hundreds of deaths related to toxic smog in London were falsely attributed to influenza.[53] The long-term public health consequences were even more serious.

The brown, noxious haze that perpetually hung over Los Angeles exemplified a new kind of pollution that was unlike the seasonal black smoke that had accompanied coal industries. The environmental cost of the massive combustion of hydrocarbons in industries and automobiles was now palpable, endangering the health of children, causing nausea, wheezing, and irritation for many, and presenting a general public health crisis to all. Special regulations would be enacted to reduce smog and ease the crisis in LA; but the broader problem persisted. Indeed, ozone levels in Los Angeles have not been below the federal health standard a single year since 1955.[54]

Nevertheless, a faith in progress, technology, and the human capacity to reshape nature to our liking persisted and created environmental catastrophes. The massive use of DDT (dichlorodiphenyltrichloroethane) as a pesticide in the 1940s and 1950s is a clear example. Strongly promoted by chemical producers, DDT became tremendously popular in agriculture, public facilities, and household use around the world. It was used on soldiers in World War II, extensively used in agriculture, and even sprayed on suburban neighborhoods to control backyard pests. As bugs developed immunity to the toxin, more and more of it had to be used. The effect on the environment was disastrous. DDT contaminated rivers and lakes around the world, poisoning the food chain, and led to the build-up of toxins in wildlife. Bird populations were decimated. American bald eagle populations crashed as DDT destroyed their capacity to hatch chicks. After a major public-relations debacle sparked by the publication of Rachel Carson's *Silent Spring*, the use of DDT was restricted; but the attitudes and values by corporate America that the promotion of DDT exemplified continued unabated. As Carson noted, 'It is an era dominated by industry, in which the right to make a dollar at whatever cost is seldom challenged. When the public protest, confronted with some obvious evidence of damaging results of pesticide applications, it is fed little tranquilizer pills of half truth.'[55]

Indeed, the war against nature continued. DDT was only one example of a new form of chemicals that dramatically increased the environmental cost of pollution in the post-war years. Synthetic chemicals such as those found in detergents, fertilizers, pesticides, and plastics increased several hundred per cent in the half-century following World War II. Once again, this reflected a growth in capital- and energy-intensive alternatives to traditional practices. The ecological contamination of these chemicals outpaced even the significant rise in industrial pollution from fossil fuels and industrial waste. Because they were not biodegradable, they migrated widely through water sources and the air, quickly becoming persistent threats around the globe. Highly toxic and often carcinogenic, these synthetic chemicals endure in air and water far longer than previous pollutants. As our daily exposure to pesticides and other pollutants increased markedly, the threat to public health, and particularly the health of children, grew grave.[56] Yet, despite devastating environmental consequences, industries promoted – and continue to promote – such synthetics with little regard for their long-term consequences. While environmental and human costs soared, so did profits; and the eyes of corporate America were fixated on the latter.

Despite important variations, in many ways the ecological impact of the competing Communist and capitalist worlds were similar. In both orders society was organized around the extraction of resources from nature for maximum profit. As the centrally controlled, highly resource-intensive, inefficient and massive heavy industries of the Soviet system struggled to keep up with the capitalist world, the environment and public health paid a dear price. Environmental laws were either nonexistent or unenforced. No regulatory agency or meaningful environmental protection policies existed.[57] In fact, since most land and resources were publicly owned and managed, the government was almost always the cause of environmental destruction. Air around most major cities grew toxic.[58] A distressing example was the massive Lenin Steelworks, which contaminated more than 4,000 square miles with its emissions. Local residents could taste the sulfur in the air. Children in the region suffered debilitating diseases at an astounding rate. Across the Soviet Union and Eastern Europe, rivers ran brown, black and green with fuel, oil, sulfur, and other industrial waste, so heavily laden with chemical pollutants they were often unable to freeze and could sustain a flame.[59] The pristine Siberian Arctic was overrun with immense mining and smelting operations. Deforestation rates rivaled the destruction of the Amazon, with over 5 million acres destroyed every year. As though to offer icing on this noxious cake, in 1986 the Chernobyl nuclear power plant blew up, leading to thousands of deaths and releasing radioactive material that would affect almost the entire northern hemisphere.

The global power play that defined the Cold War rendered the protection of the environment an irrelevant side issue. Scientific research was fixed on military applications and associated national rivalries; so research on climate change and other looming environmental crises was marginalized.[60] Protecting the natural environment, it was argued, would have to take a distant second place to winning the Cold War. For example, between 1945 and 1992, the US conducted 1,054 nuclear tests and carried out two nuclear attacks. The Soviet Union conducted its share as well. As a result, roughly 500 megatons of nuclear yield were detonated in the atmosphere. The fallout placed Americans and people around the world at an elevated risk of cancer.[61] Indeed, about 120,000 extra cases of thyroid cancer every year can be attributed to this. The test sites themselves were obliterated; and serious hazards around nuclear test sites in the Pacific, in the American southwest, and in parts of the former Soviet Union will linger for centuries.

From the perspective of poor people around the world, the Cold War battle of good versus evil looked very much like the struggle of competing empires over their fate. The US did not shy away from violence to ensure the continued security and profitability of its imperial grip. Guatemala provides a striking example. Elections in 1951 brought to power the social reformer Jacobo Arbenz Guzmán, who promised to improve the lot of the desperately poor in the country. However, when the newly elected government appropriated roughly 234,000 acres of land that had remained unused by the United Fruit Company to provide land for nearly 100,000 impoverished, landless peasants, the US took a sharp interest. The Guatemalan government had offered compensation equal to the value of the land that the American firm had declared for tax purposes. The United Fruit Company, owning half the farmable land and nearly all the railroads in the country, demanded twenty-five times that amount. With American corporate interests at stake, and many in the US administration with close ties to the United Fruit Company, the US Central Intelligence Agency (CIA) recruited an exiled Guatemalan colonel to stage a coup; and, with American support and air cover, the CIA launched a successful overthrow of the democratic government and installed a military regime. The claims against the United Fruit Company were quickly rescinded by the American-installed dictator. In the decades that followed, the US would undertake similar projects in Syria, Indonesia, Congo, Cuba, Cambodia, and twice in Chile. Justified as part of a war against the threat of Communism, in all these cases the goal was clearly not democracy, justice, or environmental protection; it was continued economic and political dominance – the further expansion and intensification of a global system that was rapidly destroying the earth's resources and impoverishing the majority of its people.

A public-private empire

Despite nominal independence for the people of Africa and Asia, the colonial system of commerce and control remained in place and ensured the impoverishment and disempowerment of the earth's majority for the benefit of European and American industries. Economies and societies in Africa, Asia, and Latin America had been redefined to suit the interests of Western industries and feed Western consumption. Although immediate political control ended with decolonization, functional control through economic constraints and political leverage remained. Land and resources continued to be in the hands of a few wealthy elites that were in alliance with or beholden to Western interests. Agriculture continued to be focused on luxury foods for the Western market such as sugar, coffee, cocoa, and fruit, or industrial materials such as cotton and rubber. Continued monoculture based on a single cash crop for Western markets continued to promote soil degradation and loss of biodiversity. Mineral resources from bauxite to diamonds were still extracted to feed Western industries, with little concern for the local population or environment.

In fact, privatized, indirect control proved more efficient than direct imperialism for Western industries. The direct political control of foreign people had proved costly; in its final decades, the British Empire was costing more money to administer and enforce that it gained in extracted resources. This more nuanced system of neo-imperialism, with control indirectly wielded through a small cadre of co-opted local elites, allowed for more efficient extraction and greater profits. Although former colonies were now supposedly independent, the same elites that had administered the country for the colonizer now ruled the country, maintaining a system of production and extraction that fed Western consumption but without the need for messy and complicated direct colonial control.

Even reform-minded leaders of impoverished countries often found that they had no option but to continue producing more products for wealthy countries. Narrow specialization made the economies of these impoverished countries particularly vulnerable. While trade from any impoverished country constituted a very small share of the total trade of a wealthy country, trade with a rich nation often constituted a large and critical source of income for an impoverished country. Roughly three-quarters of African exports and two-thirds of Latin American exports were in raw materials, minerals, metals, fuels, lumber, or other such commodities, making these countries highly vulnerable to global price fluctuations. Between 15 and 20 per cent of exports from wealthy countries were in such commodities. The result was that poor countries were far more dependent on their wealthy trading partners than vice versa.[62] Multinational corporations controlled the global network of processing, transportation, and marketing goods, leaving impoverished countries with no control over the commodities on which they were dependent. Railroads, waterways, and roads needed to remove these resources added little to the local economies and were all geared toward the delivery of valuable commodities to Western industries. Processing plants and other resources that added value to raw commodities remained outside the former colonies.

The system did what it was designed to do. The massive increase in consumption in wealthy countries was fed by this global system of extraction. The percentage of iron ore coming from the impoverished world rose more than tenfold between 1920 and 1970. Half of the world's forest loss in human history occurred between 1950 and 1990. 'Cost effective' but environmentally devastating techniques such as strip mining were used more extensively. While this activity appeared to increase economic growth in these countries, in reality the extractive industries contributed little lasting benefit to their impoverished trading partners.

With the rules of the game set in their favor, the power of corporations grew dramatically in the 1960s and 1970s. Corporations grew transnational, orchestrating a variety of enterprises and investments across dozens of countries. Foreign direct investment by US corporations increased by more than 700 per cent from 1950 to 1970. A relatively few multinational corporations control an increasingly substantial share of global commerce; and the value of the leading corporation's annual activities far exceed the total production of goods and services in many poorer countries. As a result, corporations have played an increasingly decisive role in defining global realities.

By the 1980s, global corporations increasingly sought not only raw materials from the impoverished world but also the benefit of populations so desperate that they would work for a fraction of a living wage. Factories, rather than only mines and farms, were established in impoverished countries. In the US this was called 'deindustrialization' – as industries moved south, the US economy focused more on services, sales, and information management. Economic theorists claimed that this would 'modernize' poor countries, creating a Western-style work ethic, transportation and communications networks, and a trained labor force that could lead the way to greater development. However, in reality, these industries offered little durable benefit to the impoverished. Profits were brought to shareholders in the West rather than invested locally. Political influence and financial manipulations, such as transfer pricing that set prices for transactions from one portion of a corporation to another below market prices, were increasingly used to avoid paying local taxes and evade limits on the profits a corporation can remove from the country. Indeed, the operations of multinational corporations often had little linkage with the local economy, other than the exploitation of low-wage labor; and, unlike direct imperialism, regions that did not offer profit could simply be abandoned.

The very structure of the global system ensured that any initiative would be to the benefit of 'first world' industries. This was even true of the so-called green revolution of the 1960s and 1970s, a project explicitly justified as an attempt to reduce the extreme hunger around the world. Argued to be more efficient and promising higher yields, Western methods of agricultural production were transplanted to the impoverished world. The displacement of local crops with crops for export to the wealthy countries intensified. Small farmers around the world were overcome by the consolidation of large agricultural industries. Indeed, in the absence of a change to the global economic system, the increased production brought on by the green revolution increased the benefits to the rich much more than it changed the condition of the poor.

This should have been no surprise. The cause of hunger and famine was not a shortage of food itself; it was the inequitable distribution of food.[63] In the mid-1970s, when the region that is now Bangladesh was struck by famine, this point was clear. The harvest was good and food was plentiful. However, agricultural workers had lost crucial income because they could not work during a large-scale flood; and so they simply could not afford to buy food. The dramatic famine in Ethiopia in the same period had similar causes. While there were some regional food shortages, in the country as a whole food was in ample supply. However, peasants who had been evicted from their land to allow for cultivation for commercial export could no longer afford to purchase this food. It was not scarcity of food, it was scarcity of wealth that led to these and other famines.

This industrialization of agriculture had already marked the beginning of the end for many small farmers in the United States. As large-scale, mechanized production methods and the use of massive synthetic pesticides and chemical fertilizers were transplanted to the impoverished world, a similar outcome resulted. Production costs increased, driving small farmers out of the market. Farmers had used labor-intensive traditional practices to grow diverse and sustainable yields. However, when faced with the oversupply and declining crop profits brought on by mechanized corporate agriculture, small farms were quickly displaced by large-scale operations that produced the harvest for Western markets. Around the impoverished world, tenant farmers were evicted, workers were displaced, and small-scale herders were excluded, as land was increasingly dedicated to production of sugar, coffee, cocoa, and beef for the wealthy overseas rather than staple crops for the poor at home.

This revolution in the world's agriculture had serious environmental costs. Since Western scientists pushed a one-size-fits-all approach to agriculture, traditional varieties of grains and other crops were replaced in favor of a few high-profit crops. As a result, the potentially useful characteristics of native crops – for example, resistance to local diseases or drought – were lost. Moreover, these crops required huge amounts of water and fertilizer. As small farms were replaced with large, chemical- and water-intensive farming, so soil erosion, the depletion of water sources, and chemical pollution increased notably. Use of fertilizers had more than tripled in Europe and the United States; but the increase was even more dramatic in many impoverished countries. In fact, world consumption of fertilizer increased more than tenfold from 1950 to 1990, to 145 million tonnes each year. Moreover, because these new crops lacked the natural immunities of traditional crops, they required massive quantities of pesticides. The hard currencies needed to purchase these pesticides and fertilizers made impoverished economies even more dependent on the terms of trade set by the wealthy.

The green revolution was not, in the end, a real effort to end hunger; it was an attempt to circumvent the need and demands for meaningful reform in the rules of the global game by promising more food for more people.[64] Delivering on this promise would have required deep and broad-based changes in the global structure of food production and distribution; both industrial leaders and small farmers knew that was not on the cards. Still, the green revolution did increase crop yields notably, which, it was expected, would reduce world hunger. And, in fact, the percentage of malnourished people fell by about 20 per cent between 1965 and 1990. However, with an increased global population, about the same number – over 800 million people – remain undernourished. Large landholders used their land to produce crops for export to the wealthy countries of the world rather than to feed those who had no money to pay. Countries that had been self-sufficient in food production a few decades previously became net food importers. Small peasant farmers could not make the kinds of investment necessary to adopt these new practices and were largely displaced by large corporate farms. As small landholders became landless laborers, the gap between the wealthy few and the impoverished majority increased. So, overall, the green revolution did not challenge the fundamental inequity of the global system; but it did notably increase the environmental and social damage this system wrought.

Agriculture simply reflected the global system of production more generally. Export for profit took precedence over the welfare of the people. Multinational corporations manipulated local manufacturing and trade, influenced governments, evaded taxes, exploited labor, and brought their profits back to banks and investments in wealthy countries. Elites in impoverished countries who collaborated with multinationals enriched themselves, leading to dramatic inequalities within and between countries. The disparity between the world's wealthy few and the impoverished many had unquestionably grown. Little or no income gap had existed between Europe and the rest of the world before the Industrial Revolution.[65] By 1945, for each dollar earned by a person in an impoverished country, a person in a rich country earned twenty. By 1975, that ratio was forty to one.[66] While some impoverished countries did marginally better, many stagnated and quite a few suffered declining income and living standards.[67] Within impoverished countries, the gap between the rich and the poor was often much greater than in wealthier countries.[68] This growing inequality was accompanied by cultural degradation and political repression.

Corporate globalization

By the early 1970s the combination of huge government spending and high public consumption placed serious strains on the US economy. The 1950s and 1960s marked a significant increase in peacetime military spending and a persistent balance-of-payments deficit. As early as 1953, military spending accounted for 10 per cent of all goods and services produced in the US.[69] By the 1970s US interventions around the world, particularly the sustained venture in Vietnam, had grown quite costly. More generally, the US was consuming far more than it was producing. From 1960 to 1970, the US went from a nearly $5 billion trade surplus to over a $2 billion trade deficit.[70] Having bought much around the world on credit, policymakers now had a problem: the amount of foreign-held dollars was greater than the amount of gold the US had to back up the value of the dollars. The basic rules of the international system would have required the US to adopt the same harsh austerity measures it had forced on other countries. Some belt-tightening was in order.

President Nixon had another idea. In 1971 he announced an end to the US commitment to redeem dollars for gold, and imposed wage controls, a surcharge on US imports, and a depreciation of the dollar. The move was clever, and would only have worked for the world's leading economy. The US dollar's value had been assured by a fixed gold standard after World War II; and an international system of fixed exchange rates tied to the dollar allowed other countries to stabilize their own currencies with the US dollar. With Nixon's move, the dollar was no longer backed by gold; but the world's countries had no choice but to maintain the value of the currency, since they held substantial amounts of US dollars in their own reserves and were dependent on a stable exchange rate for their own economic well-being. The US had essentially exported much of the cost of its war in Southeast Asia and general overconsumption to Europe and the rest of the world.

With their privileged position in the world system, American industries and consumers could continue their massive consumption uninterrupted. High interest rates kept the value of the US dollar high, allowing Americans to buy more foreign products for less money. Normally, this would be a problem since it would also make it difficult to sell US products overseas; but since the US was willing and able to run large budget and trade deficits and take on massive debt, this problem was not as acute as it would have been for a less powerful country. The US spent tens of billions of dollars on imported oil each year, and ran an increasing trade deficit. By the 1990s, the US would be consuming over $100 billion more than it was producing each year. US military spending jumped more than 50 per cent in the 1980s, with spending topping $300 billion each year by the end of the decade, an increase funded by a huge debt. With the global system centered on the US dollar and commerce, the game was set in the US's favor; it could afford to increase consumption since the rest of the world would share the bill.

Meanwhile, distracted by its war in Vietnam, the US had allowed its military control in the Middle East to wane, weakening its control over the world's oil supply. Middle Eastern oil had become more critical to US industries. In fact, domestic oil production had declined from 43 per cent of global consumption in 1957 to 21 per cent in 1972, while oil imports to the US rose sharply.[71] Oil-producing countries, recognizing this shift, challenged the control of oil production and pricing by Western corporations and began demanding a greater share of the profit from the oil pumped from their countries. When the US backed Israel in its war with Egypt, oil-producing countries that were hostile to Israel acted in concert to push prices up dramatically. The tremendous dependence of the world system on oil was reflected in the deep economic slowdown brought on by this price rise. Yet, in spite of the price increase, the short-term demand for oil did not drop significantly.

The most severe effects of the oil crisis were felt in the world's impoverished countries, as the resulting need for hard currency to pay for mounting oil prices intensified a mounting budget and debt crisis. The money sent to elites in oil-producing countries to pay for the massive amounts of oil consumed by wealthy countries had found its way to Western banks in the 1970s and 1980s; and these banks, hoping to reap greater profits, sought out borrowers in sometimes-desperate and often-nefarious leaders of impoverished countries. In a fluster of predatory lending, loans were offered faster than any country could reasonably hope to manage the debt. Often the money was used to pay for development projects, or to lower the cost of food and fuel for the population. At times, however, corrupt leaders enriched themselves with the help of Western banks, placing their countries in deep debt and slipping the money into personal overseas accounts.

Many countries might initially have had a good shot at paying their debt burden; however, when oil prices rose again in the late 1970s, poor countries were hit hard. Rising oil prices meant they needed to spend more of their budget on oil imports. At the same time, the prices of the commodities they sold to the rich dropped, reducing their revenues sharply. Most important was the effect of rising US interest rates. The US pushed interest rates to unprecedented highs to ensure a strong dollar, thus

2.8 / Debt of heavily indebted countries in early 1990s

Total debt ($)		As % of exports		As % of GNP	
Brazil	121,110	Guinea-Bissau	6,414	Nicaragua	750
Mexico	113,378	Nicaragua	3,162	Mozambique	495
India	76,983	Sudan	2,961	Tanzania	178
Argentina	67,569	Mozambique	994	Jordan	163
Thailand	39,424	Argentina	450	Jamaica	132

Source: World Bank, *World Development Report, 1994*, Oxford University Press, New York, 1994.

helping cover the cost of massive US military spending and oil imports. This raised the effective amount of the impoverished world's debt, since the debt needed to be paid back in now-more-costly US dollars. And, since the interest rate of these loans was tied to US interest rates, their interest payments soared. Countries took on more debt in an effort to keep making their debt payments. Often, as a condition of continuing support, governments were required to assume responsibility for private debts incurred in their country. Much of this debt had in fact been acquired by subsidiaries of multinational industries operating in their countries. So the fiscal burden of impoverished countries was at times significantly increased when they were required to take on the debt of Western corporations!

The effect was dramatic. Poor countries, now saddled with a huge debt, were unable to import even the most basic goods and became more vulnerable to changes in commodity prices. It was not simply the economies of the impoverished that were crippled by this global system of exploitation; the capacities of poor countries to define their options, identify visions of hope, and empower their people were thoroughly undercut. Many countries owed far more than they could hope to pay back. Peru, for example, owed $14 billion. Just its yearly payments to keep up with the debt exceeded the total value of all exports from Peru each year! Far more money was now being paid by the world's poor countries to pay their debt than was being received in development aid. In desperation, poor countries continually increased their export of raw materials to pay their debt, with their forests, soils, and waterways paying a high price.

The hardship felt by the American economy brought on by the oil shock, on the other hand, was comparatively minor and short-lived. Though oil prices rose significantly, the supply of low-cost energy was relatively

2.9 / Mounting debt (foreign debt of less developed countries and former Eastern bloc, $ billion)

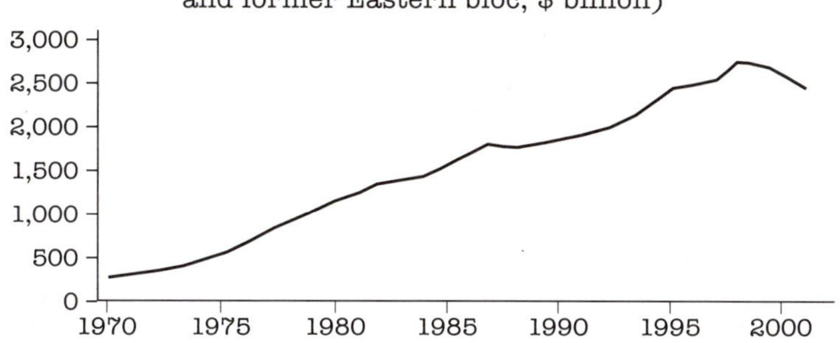

Source: World Bank, *World Development Report, 1994*, Oxford University Press, New York, 1994; Michael Renner et al., *Vital Signs 2003*, Worldwatch Institute, Washington DC, 2003.

quickly re-established through the use of coal and the development and expansion of alternative oil sources in Mexico, the North Sea, and Alaska. Divisions within oil-producing states led to the collapse of solidarity and lower crude prices. In fact, by the late 1980s, real prices for oil and gasoline would be lower than they had been before the oil crisis.[72]

American consumption continued to skyrocket in the 1970s. Disposability grew more common as Americans brought products with a useful life often measured in days if not minutes. Cheap paper and plastic products flooded the market. Cars made in the 1970s had a useful road life far shorter than those made two decades previously. Energy demands grew sharply with this insatiable consumption. When hydroelectric plants could no longer keep up, other sources were exploited. Nuclear energy was tried, but the political, and environmental consequences did not outweigh the disappointing economic benefits. Necessary safety measures meant that it was more expensive than initially promised; and the problem of disposing of waste with a dangerous life in thousands of years proved politically difficult. Coal-fired plants were relied on heavily. As smokestacks were built higher to hide the toxic effects of the plants from the local populations, these plants spewed acidifying contaminants high into the atmosphere across state and national borders, degrading forests and increasing global warming.

So, while at first glance it might seem that the solidarity demonstrated by oil-producing states and the resulting oil crisis underscored the vulnerability of the world's wealthy countries, in fact the opposite was the case. The oil embargo of the 1970s made it clear that a disciplined international effort by an organized and motivated association of countries that control the vast majority of the most crucial commodity for Western industries and lifestyles – oil – could not mount even a temporary barrier to continued

Western expansion and control. The oil crisis did not significantly change lifestyles or consumption patterns in the world's wealthy countries. It didn't even really change the availability of oil to Western consumers for very long. The real burden, in some ways the only significant burden, was wrought on the world's impoverished countries that were caught in the backwash of US consumption.

By the mid-1970s the global system of exploitation appeared unchallengeable. With decolonization in the 1950s and 1960s, a wave of opposition spread throughout the impoverished world. Leaders demanding a more just international order and promising reform arose in Africa and Latin America. Some were virtuous, some were not. It didn't matter much in the long run, since in the end the dominance of US and European interests would prove far harder to overcome than these new leaders had imagined. A clear example was offered in 1974 when, after more than a decade of collective work, a group of impoverished countries passed a united declaration in the UN General Assembly calling for fundamental changes in the global economic system. This call for a so-called New International Economic Order (NIEO) outlined changes in the international distribution of wealth, decision-making power in the World Bank and International Monetary Fund (IMF), and changes in the terms of trade that would allow impoverished countries an opportunity to develop. The United States and its wealthy allies were unimpressed. Despite minor conciliatory lip service, no real change took place. The impoverished countries of 'the South' simply had no power to force their demands in a global system deeply entrenched to the benefit of the wealthy countries of 'the North.' Any serious effort by impoverished countries to force the North to compromise – repudiation of debt to Northern banks, nationalization of assets of Northern corporations, cartels to control the prices of resources sold to global industries – would hurt the impoverished countries far more than it would hurt the North. Despite the promise and enthusiasm it had held for many in the impoverished world, the proposal for a new international order died a quiet and ineffectual death.

With a system of control so complete, enforcement through direct violence was relatively rarely needed. However, when the demands for reform by the impoverished could not be put down by intimidation and economic pressure, Western governments were ready to use violence. In 1983, for example, the US funded and directed a war on the reformist Sandinista government of Nicaragua, ending a successful environmental protection program, causing over $4 billion in economic and environmental damage, and costing the lives of 60,000 Nicaraguans. The US Congress itself uncovered clear evidence that the US-backed forces were involved in rape, torture, and horrific human rights violations; but, clearly, ensuring respect for human rights was not the goal.

Most of the time, no violence was needed to enforce control. For example, in the mid-1970s, when the moderate reformer Michael Manley was elected to power in Jamaica, the pressure wielded through the system itself was enough to end his efforts toward improvement. Manley imposed a levy on bauxite – a major American import used to make aluminum – to fund education and social development programs in the country. The programs had mixed initial results, but they enjoyed a great deal of popular support. In a bold move, this new levy was tied to actual market prices, rather than to the arbitrary value American corporations assigned to the commodity. American firms, fearing that their complete control of the resource might be jeopardized and furious over losing part of their considerable profits, punished the Manley government by shifting to African sources and cutting back on their production of bauxite and alumina in Jamaica. This undermined the government's revenue hopes and sparked increased unemployment and strikes. A press campaign by the US industries blamed Jamaica for the rising cost of aluminum, even though corporate profits at the time enjoyed a boost. To add to the pressure, the US government embargoed aid to the country and wielded political pressure, covert and overt, to destabilize the Manley administration.[73] Finally, the Manley government buckled, and a government friendly to American corporate interests came to power. Having learned his lesson, Manley later re-emerged on the political scene as a clear supporter of friendly ties with the United States and its industries.

While an organized resistance of the oppressed was unlikely to challenge the global dominance of Western governments, a transfer of power *was* taking place. Power was not moving into more democratic public hands; it was moving increasingly into private hands, into the hands of expanding multinational corporations. The annual sales of dozens of corporations reached the tens of billions of dollars. The largest had more economic activity than the total goods and services of three-quarters of the world's countries. The influence of these industrial behemoths reached around the globe, into the largest countries and the tiniest villages. The amount that they spent building factories, mines, and other enterprises around the world began to increase at an incredible rate and would exceed $1 trillion by the end of the century. With this tremendous influence and control over resources, multinational corporations increasingly defined the lives of people and communities around the globe.

At the same time, with the end of the gold standard, a currency market emerged that placed control of the US dollar in the hands of private firms and speculators. As speculators bought and sold the world's currencies, the fates of nations climbed and plummeted, and there was relatively little any government could do. When the US tried to regain control of the money supply, US banks simply went overseas to continue operations. More generally, as private interests and financial markets gained dominance, the government's capacity to influence economic activity weakened. Overall, the tremendous expansion of multinational firms, the rise of currency markets dominated by private speculators, and the expansion of US international banking increasingly placed the economy out of the hands of democratically accountable governments and in the hands of private corporations out to make a profit. This process of corporate globalization redefined the world and its institutions to ensure corporations unimpeded access to the world's resources and unregulated markets for their products.

A wave of New Right governments facilitated this transfer of power by radically eviscerating any policies that allowed public control of the economy. The Reagan administration in the US and Margaret Thatcher's government in Britain ushered in a neoclassical economic orthodoxy that cast government controls and regulations as inherently bad and advocated complete surrender to market forces. Regulations to protect the environment were gutted. Social programs were attacked as an impediment to economic growth. And any policy that promised the further enrichment of wealthy elites was hailed as sound economic strategy. As we will see in the chapter that follows, international organizations like the World Bank and the IMF used the leverage gained from international debt to enforce this surrender to unregulated markets globally. Both these institutions worked aggressively to tear down any impediment to corporate-centered globalization.

Since the people put governments, and not corporations, in place, surrender to the market meant a loss of democratic accountability. Corporations have no meaningful national allegiance. In fact, by the mid-1980s, few products still had a clear national origin. Any car, refrigerator, or television has parts from countless countries, made from raw materials from other countries by workers who have been brought in from still other countries and with energy from fossil fuel that's extracted from yet another country. Trade occurred at virtually every stage of the production chain through a network of producers and distributors, subcontractors, and subcontractors to subcontractors. This not only meant that the social and environmental impact of any item was increasingly impossible to tally, but that corporations were neither dependent on nor loyal to any particular market, workforce, or resource base. The economies, stability, and shared fate of communities around the world grew increasingly dependent on the resulting corporate free-for-all. The freedom to shift their investments, relocate production, and transfer industries as they chose gave transnational corporations tremendous influence. Their capacity to define the rules of the global game placed corporate interests and priorities squarely at the center of the globalization process.

Beyond its damage to democracy, this expansion of global trade had inherent environmental costs. Trade involves expanded shipping of goods over longer distances; and this means more use of fossil fuels and more pollution for each item that's made.[74] Thus a head of iceberg lettuce that offers 50 or so calories of dubious nutrition might require more than 1,000 calories of energy – and the associated production of greenhouse gases – to get from the farm to the supermarket.[75] The ingredients of an average carton of yogurt, for example, travel about 1,200 miles before being mixed together to make the final product. Then that product is shipped perhaps hundreds of additional miles to the final consumer.[76] The need for refrigeration all along this path doubles the energy requirements. The increasing use of air transportation in trade massively increases these costs. A two-minute takeoff of a Boeng 747 emits as much pollution as 800,000 lawnmowers running for a full hour![77] The indirect environmental impacts are equally compelling. Goods that are shipped need to be packaged, for example, and their components are packaged and shipped as well. In fact, by the time a product finally reaches its consumer, several stages of packaging, shipping, and repackaging have probably taken place. The amount of waste is staggering.

Throughout the 1990s, American consumption continued its unabated increase. Consumer spending comprised a full two-thirds of the nation's total economic activity. Savings and investment fell while spending rose. Now that wealthy Americans had all their basic needs and desires met, spending increasingly shifted to amenities, leisure, recreation, specialized items, disposable items, and so on. To keep this consumption accelerating, marketing had grown increasingly sophisticated and insidious. Cars, appliances, televisions, clothing, food, and shelter had come to be recast as fashion amenities based in fickle social styles, and thus designed with short-term obsolescence in mind. Americans dutifully continued to increase their buying. They purchased new wardrobes each year, far before the clothes actually wore out. Functioning appliances were replaced regularly with new appliances, updated with the latest style alteration or technological amenity. A sophisticated and omnipresent system of marketing and branding made the act of shopping itself the goal. Millions of suburban teens spent their leisure time in malls, socialized to see consumption itself as a mode of entertainment and personal fulfillment.

2.10 / The rich got richer... and the poor got much poorer
(% change in average American household net worth, 1983–1998)

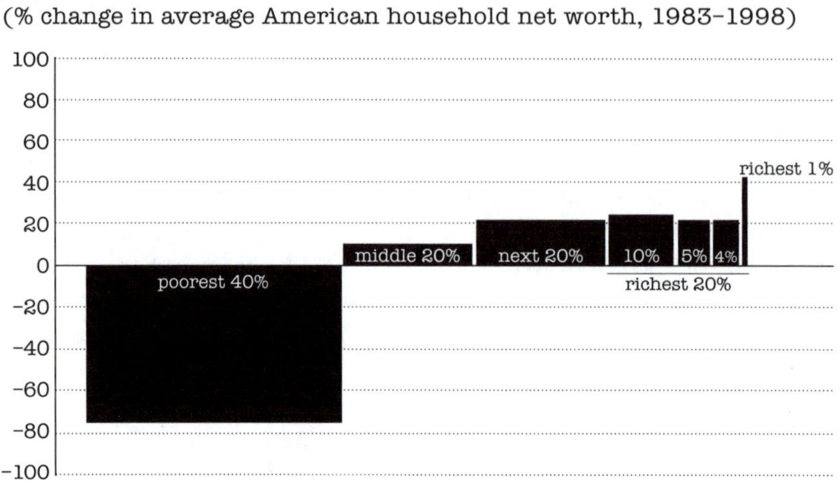

Source: Edward N. Wolff, 'Recent Trends in Wealth Ownership, 1983–1998,' Working Paper, Levy Economics Institute, April 2000, table 3, www.levy.org.

Meaningful improvements in life had been put aside in favor of accumulating more stuff. Americans hardly recognized the increasingly perverse effect on their quality of life. Despite their greater productivity, Americans worked harder and longer than before. As leisure time plummeted, rates of work-related stress and illness rose.[78] The American Dream turned into one of those nightmares where you run faster and faster but can't move forward.[79] Inundated by marketing in virtually every aspect of their daily lives, convinced that fulfillment could be found in buying more cheap stuff, and always aware of the ever-present threat of real poverty, Americans stayed on the treadmill, working more and more so they could buy more and more. As they increasingly narrowed their lives to a cycle of work and consumption, personal debt skyrocketed and their life satisfaction dropped.[80] In the 1990s, each American spent more than $21,000 per year on consumer goods.[81] An all-consuming epidemic, a 'painful, contagious, socially-transmitted condition of overload, debt, anxiety and waste resulting from the dogged pursuit of more,' defined the American society and economy as well as the national psyche.[82]

Meanwhile, the division between the rich and poor in the US grew dramatically. The richest 1 per cent of Americans saw their income soar from $420,000 in 1979 to $1.2 million in 1997, a shift facilitated by more than a decade of corporate-friendly policy reforms. In this same period, the wages of the majority of Americans actually dropped. Overall wealth grew even more concentrated than income. By the end of the century, the wealthiest 1 per cent of households owned nearly 40 per cent of the country's wealth. The bottom 80 per cent owned just 17 per cent of the nation's assets. Hunger and homelessness grew without interruption in the 1990s; and, a full 17 per cent of Americans lived below the poverty line. In the end, the corporate-defined, free market policies that were wreaking havoc in impoverished countries were also leading to the impoverishment of many Americans.

A globalized production scheme allowed for increasing distance between the consumer and the social and environmental impact of what was being consumed. A product consumed in California, for example, may have a serious and lasting public health, social, or environmental impact on communities and ecosystems in India. But consumers in California can comfortably ignore (willfully or otherwise) the real costs of their consumption in a way that they simply could not do if the consequence were being felt in their own backyards. Even concerned consumers found it increasingly difficult to make informed choices. If the product is made half a world away, the environmental or social costs entailed in its construction may not be clear. Ignorance is bliss. Americans could continue to call themselves environmentalist as they drove their SUVs. The consequences of their exhaust will be felt first in places halfway around the world. Shoppers could happily bargain-hunt for their sweatshop-produced clothes, made comfortable by the calculated ignorance of the real conditions in which the clothes were made.

No one person designed this system. No single maniacal group of corporate masterminds planned out an international system of exploitation that would impoverish the many and destroy the earth. But the system could be no more absolute if they had. The rules of the game are entrenched in laws and enforced with violence; the priorities of the key players – profit and growth – are justified by a claim that if each of us pursues our narrow self-interests, everyone will benefit. Experience shows that nothing could be further from the truth. The tremendous wealth of the few has been built on the impoverishment of the many. As the impoverished world chokes on the exhaust of our wealth, as our gluttony intensifies world hunger, as our rabid consumption destroys the atmosphere and carries us closer to the abyss, we might keep a simple observation in mind: this system was made by humans, and it can be changed by humans.

Making the rules / 3

As we careen toward environmental catastrophe, we've got to know that this isn't the only way things could be. The current pattern of global injustice and environmental destruction is not foreordained or the result of some irrevocable law of nature. There is nothing about human beings that makes social exploitation or environmentally devastating waste inevitable. This global system of destruction isn't set in the rules of nature; it's imposed by the rules of people. So, the question arises, who makes these rules? How are they enforced? More to the point, who the hell's in charge here?

Well, in some sense, we all make the rules. Or, at least, we all follow them. We buy toxic products, drive polluting vehicles, and benefit from the poverty of others as a normal part of our everyday lives; and, through our participation, we all share some culpability in the way things work. But that doesn't mean we all have equal power. And it certainly doesn't mean that no one's in control. In fact, the current scheme of corporate globalization is all about control. It's about maintaining control in the hands of a wealthy and powerful few.

A web of international institutions is making sure that this global corporate takeover is complete and that any barriers to corporate access and control are eradicated. The International Monetary Fund and the World Bank define the options for the world's impoverished countries. They are the global enforcers of the free-market economic orthodoxies;

and, together, they help ensure the current system against challenges from the downtrodden. The North American Free Trade Agreement and other regional trade agreements further empower corporate interests, giving them the ability to strike down environmental and social protections that in any way limit their profits or restrict their capacity to exploit precious natural resources. The World Trade Organization, the great colossus of international institutions, is redefining the world order to guarantee their access to even the smallest corners of the globe. Human rights, vital ecosystems, social justice, and a generally livable planet are trampled by a determination to redefine the world for corporate interests.

And so sovereignty, the right of a nation to determine its own priorities and fate, and democracy, the right of the people to govern their own lives, are pushed aside. After all, with sovereignty goes democracy: if governments are powerless, or effectively controlled by powerful corporate interests, then even the most virtuous public officials are made powerless; and the people are made powerless as a result. It may appear that in wealthy countries we've held on to a greater ability to define our own laws and priorities, but this is a fading illusion. The ever-tightening collusion of corporate interests and government leaders that are beholden to them is placing our collective fate increasingly in the hands of the unelected.

In fact, the line between the elected and the unelected is not at all clear. Public and private elites slip back and forth between leadership positions in government, corporations, and financial institutions. Powerful positions in government hinge on the support of private interests; and private interests depend on their alliance with public policymakers to redefine the world to their favor. Under the explicit pursuit of free trade, these interests work to strike down environmental and social regulations, open access to the farthest corners of the world to exploitation of precious resources, expand and control a singular system of production and trade, and generally tear down the resistance of public leaders, activists, and others around the world who favor a sustainable and just future.

The result can be called a new imperialism. Its scope, power, and reach are increasingly total; at its helm are the world's wealthiest and most powerful corporate and government leaders; and its rules are embodied in a coordinated array of international institutions that hold the free market as sacrosanct. The resulting system is complete, expansive, and every bit as devastating as past systems of imperialism. But this system is enforced with a neatness, formality, and executive sterility that makes it appear legitimate and proper. This highly inequitable, socially devastating, and environmentally unsustainable global order appears to most of us as the normal, matter-of-fact way the world works. It seems both natural and reasonable. But it's not. It is designed and enforced by an entrenched web of unaccountable institutions.

Banking on poverty

The World Bank and the IMF have become icons of corporate globalization. Sharply under the control of the world's wealthiest countries (and their corporate sponsors), these institutions have championed the notion that the best way to ensure a better future is to give corporations a free hand and do absolutely nothing to challenge the status quo. You will remember from the previous chapter that when it was first started after World War II, the IMF's chief aim was to help stabilize a global system of fixed currency exchange rates. The IMF's job was to provide funds for countries whose economies were in a downturn so that their currency didn't lose too much value. But also, as you will recall, when the US ran up a major deficit, President Nixon took the country off the gold standard. The not-unintentional effect was to have Europe and Japan share some of the cost of US exploits in Southeast Asia.

The resulting end of the international system of fixed exchange rates could have meant the end for the IMF. After all, since the IMF was established in large part to support a system of stable exchange rates, it was now left largely without a mission. But the domination of the world's wealthy countries seemed to be coming under threat from the early success of OPEC, the demand for a New International Economic Order, and other alliances and efforts by the world's impoverished to challenge the global status quo. So the IMF was given a new job: ensure the security of the global status quo in the face of challenges. While espousing the virtues and infallibility of unregulated markets and corporate expansion, the IMF began to take a leading role in defining the fate of impoverished countries.

The rising debt of impoverished countries gave the institution the leverage that it needed to impose its demands. The system of international debt that crystallized in the 1970s and 1980s as an indirect outcome of mounting oil prices crippled the economies of impoverished countries and empowered the IMF, western governments, and international corporations to ensure that real change in the global distribution of wealth remained unattainable. The IMF's aim was not simply to ensure debt repayment, though this was always a priority, but to use this debt dependence to impose deregulation and corporate-friendly policy changes on the world's most impoverished countries. At home, the Reagan administration was busily destroying environmental protections, undercutting social protections for workers and the poor, and assuring a free hand for corporations. The IMF would make this transformation global.

Thus, while the IMF was originally conceived in part as a deficit financer (a source of needed funds when countries ran budget deficits so they could continue to stimulate economic activity), now it adopted exactly the opposite approach.[1] In the face of an economic downturn, the IMF demanded serious belt-tightening, never mind that economic logic indicated that cuts in spending could in fact increase the downturn and feed a spiral of economic decline. Because the IMF could cut off needed funds, and, more crucially, discourage investment from private interests, most impoverished countries were cowed into submission. The result, unsurprisingly, was twofold: under the IMF's programs, impoverished countries almost always experienced continued economic downturn; and international corporations enjoyed increasingly unrestricted access to resources and markets around the world.

This new role for the IMF was coupled with a redefined role for the World Bank. Initially established as a marginal player in the European post-World War II recovery, the World Bank's enhanced status was sparked by a proposal from impoverished countries in the late 1950s for a Special United Nations Fund for Development (SUNFED). The US and other wealthy countries rejected the idea of an organization under the democratic control of the UN with each country having an equal vote, and moved to offset the demand for SUNFED with an agency that would be firmly under the control of wealthy countries with voting based on the size of a country's contribution to the agency's bank account. In short, the more wealthy the country, the more control it would have. And so the World Bank's International Development Agency was born.

Like the IMF, the aims of the Bank were to ensure the stability of the global system and promote the interest of Western industries. In the wake of the oil crisis and the failed demand for a so-called New International Economic Order, in the early 1970s, Robert McNamara, the former Secretary of Defense of the United States, was given control of the Bank, and, in the space of a decade, increased its yearly lending from $2.7 billion to $12 billion.[2] Projects were guided by government or military elites and imposed with little input, much less participation, from the local population. Cold War priorities dominated US policymaking; and supporting anti-Soviet regimes trumped any commitment to development or democracy, much less the environment. Bank projects supported brutal repression in Brazil, the Philippines, Indonesia and many other countries. In fact, World Bank support helped prop up some of the world's most vicious dictators. Although the Bank has now moved away from such blatant support of tyranny, the general pattern has not changed.

As a recent United States Department of the Treasury report has it: 'Since the founding of the World Bank in 1945, we have been their largest and most influential contributing member. We have also been their largest beneficiary in terms of contracts awarded to U.S. firms.'[3] More generally, the Bank ensures that global investments and development support US interests. For example, after the oil shocks of the 1970s, US policymakers, concerned with dependence on oil imports from OPEC countries, pushed the Bank to promote fossil-fuel exploration and development to guarantee the US diversified sources of oil.[4] The results not only weakened OPEC's control of oil prices but forced the sale of state-owned oil and gas industries around the world to Western corporations and their subsidiaries. Despite the fact that this pushed energy costs further out of reach for impoverished populations around the world, the US got what it wanted: a diverse supply of low-cost energy for Western industries and consumers.

The World Bank and the IMF are among the most powerful institutions in the world, and they're run by an old boys' club. Both are controlled by the world's wealthy countries, defined by Western industrial and financial interests, and sharply constrained by a US veto. Although the Bank's focus is on impoverished countries, it is always headed by an American; and the IMF, by tacit agreement, is always headed by a European. Leaders are chosen behind closed doors, with no required input from the countries for which they will make policy, and with no required experience in international development.[5] Given that they are isolated from public scrutiny, and in no way democratically accountable, one should expect nothing else from these policymakers but that they serve the interest of those who hired them.

The result is that the options for impoverished countries are defined and limited by the so-called 'Washington Consensus' – a broad agreement between IMF, World Bank, and US policymakers that rests on a rabid commitment to free-market economics. With the leverage provided by the global debt crisis, these institutions enforce the dictates of modernization theory, arguing that only full exposure to global markets will lead to development. So-called structural adjustment programs (SAPs) imposed on poor, indebted countries by the IMF require the opening of their markets to global competition, drastically reducing government regulation, ending price supports on energy, food, or other items, forgoing wage controls, eschewing any policy that might inhibit corporate profit, and, in short, yielding completely to the demands of international industries. Buckled under the strain of mounting debt, and with the terms of trade and finance stacked against them, impoverished countries have had little choice. At IMF insistence, they change their laws to allow foreign companies to buy up crucial industries at bargain prices; they enact deep cuts in spending on education, health care, social programs, development projects, and food programs; and they eradicate barriers that might inhibit the invisible hand of corporate control.

The hypocrisy is thick. While the US and Western Europe maintain high subsidies for agriculture, such subsidies in impoverished countries are targeted by Bank and IMF officials as barriers to trade. While the IMF insists that impoverished countries open their markets to Western exports and investment, Western markets are often closed to the products of impoverished countries. While the US continues to use tariffs to protect vital industries, impoverished countries are forced to eradicate all such supports for their own industry.

When coupled with the stress of the debt crisis, declining commodity prices, and deteriorating terms of trade, these imposed policies have been catastrophic. Over the course of the 1980s and 1990s, when these so-called structural adjustment programs were in full swing, poverty deepened and inequality worsened. Rising poverty meant that the number of undernourished people grew. Public health care systems collapsed. Women were hit especially hard, as they often absorb the brunt of economic and social decline.

There are any number of examples, each following a familiar pattern. Under pressure from the IMF, countries are compelled to refocus production and economic activity on exports. Agricultural production might be shifted to luxury crops – exotic fruits, for example – for American and European consumers. The required changes are often financially and environmentally costly. Large-scale dam and irrigation projects might be required to feed the new industries, for instance. Most often, large landowners reap the greatest benefit while small-scale farmers feel the greatest pressure. The debt incurred to pay for these projects rises sharply. In many cases, national debt exceeds the total value of all goods and services in the country. Because these reforms mean a shift away from self-sufficiency in food production toward dependence on the international market, rising food imports are common. Government food subsidies that may have kept food at affordable prices are reduced or eliminated at IMF insistence. This can push wages up and reduce a country's competitiveness. More crucially, food grows out of the reach of the poor. Price subsidies for large producers of export crops, however, are often not cut. So wealthy landholders are less likely to suffer. In the end, the amount a country spends servicing its debt can be many times what it spends on education or healthcare.

Even where appropriate, establishing a functioning market-based system takes greater care and more time than the IMF is often willing to invest. With a fixation on market liberalization and privatization at all costs, IMF officials have pushed for virtually immediate liberalization, wreaking tremendous hardship on the economies of impoverished countries. Joseph Stiglitz, former chief economist and senior vice-president of the World Bank, described the IMF's repeated policies of 'forcing liberalization before safety nets were put in place, before there was an adequate regulatory framework, before the countries could withstand the adverse consequences; forcing policies that led to job destruction before the essentials for job creation were in place; forcing privatization before there were adequate competition and regulatory frameworks.'[6] This blind 'market fundamentalism' called for the cutting of food subsidies before a stable market for food was in place, which has led to massive hunger, and the cutting of subsidies on fuel, so people can no longer afford the kerosene they need to cook or the fuel they need to heat their homes. No matter the circumstance, the answer for IMF officials was rapid and radical deregulation and privatization, willfully blind to the cost to the people who must endure the ensuing hardship.

Around the world, strikes, demonstrations, and even riots have broken out in protest. In Ecuador, tens of thousands opposing IMF reforms and the deepening economic crisis stormed the Congress, catalyzing a bloodless coup. In Argentina, a government was brought to its knees. In Bolivia, a president who has consistently put the interests of corporations and Western institutions above the needs of the country's people was forced to resign when hundreds of thousands took to the streets in desperation. In Brazil, more than a million people voted against further IMF reforms in a symbolic referendum.[7] Despite continued protest and unrest, the IMF persisted with its policy demands. In India, 10 million state employees went on strike, calling for an end to IMF and World Bank policies. In Nepal, Pakistan, Turkey, Mexico, Indonesia, and many other impoverished countries around the world, construction workers, small business owners, clergy, teachers, doctors, lawyers, mothers and fathers have taken to the streets to demand an end to IMF-imposed policies and the IMF takeover of their government.[8] Protests at meetings in Seattle, Washington DC, Quebec, and Geneva have been so persistent and large that the IMF, the WTO, and related organizations have taken to having their meetings in more isolated locations, from which the inevitable swarm of protesters can be barred.

By increasing inequality and poverty, IMF-imposed structural adjustments can have a devastating impact on the environment. Local, environmentally sustainable systems of subsistence are often crushed under the pressure of mounting poverty. Land-hungry peasants and farmers turn to over-cultivation, unsustainable land clearing, and other destructive practices to survive. For example, if the price of heating fuel is out of reach, the cutting of trees for cooking and warmth, an important contributor to deforestation, will increase dramatically. Under the pressure of poverty and declining security, loggers are inclined to overcut, fishers are led to overfish, and governments are unlikely to dissuade foreign investment by enforcing environmental laws.

Costa Rica provides a clear example. In the 1980s alone, Costa Rica underwent nine IMF and World Bank stabilization and structural adjustment programs. The aim, argued IMF and Bank policymakers, was greater economic 'efficiency' by opening up the country's industries to foreign investment and exposing the domestic market to imports. Two key industries, bananas and cattle – both largely foreign dominated – benefited from these policies. A combination of subsidies and tax cuts on banana exports triggered a dramatic increase in production and an equally dramatic increase in the rate of deforestation. The industry's heavy reliance on pesticide meant a serious rise in health problems for plantation workers and the local populations, as well as increased contamination of land, rivers, and eventually the ocean. Indeed, most of the destruction of the coral reef off the Costa Rican shore is linked to chemical runoff. Increased subsidies to cattle ranchers had similar effects: as beef exports increased, so did land degradation and the destruction of the forests. Yet, in the perverse logic of rabid free-market economics, these precious natural resources were an acceptable cost to pay for the sake of continuing to feed the demand for cheap beef from an expanding fast-food industry in the West.

Costa Rica is only one of many examples around the world. In Ghana in the early 1980s, for example, structural adjustments focused on increasing cocoa exports (despite, as we'll see in Chapter 5, the common use of slave labor in the industry). The resulting increase in production caused the price of cocoa to drop on the world market. This in turn led to increasing hardship as revenues plummeted. The country turned to its tropical forests to make ends meet. Timber production more than doubled and deforestation rates proceeded at nearly 2 per cent per year. Over three-quarters of the country's tropical forest is now gone, leading to soil erosion and regional climate change. Soon Ghana will actually have to import wood, making it even more dependent on the world system.

Policy changes pushed on Vietnam by the IMF and the World Bank in the 1990s have had similar consequences for that country and a broader impact on the fate of many other impoverished countries. Encouraged to open its markets and shift its focus toward generating foreign exchange, the Vietnamese government began an aggressive program to persuade farmers to shift from rice production toward the production of coffee for export. Because Vietnam is not particularly well suited to coffee production, the bulk of the crop was lower-quality robusta beans, rather than the more valuable arabica beans. Nevertheless, in ten years, Vietnam moved from having an almost negligible coffee crop to being the second largest coffee-producing country in the world.[9] When world coffee prices collapsed as a result of overproduction, the hopes for the new industry and an export-led economic miracle dimmed. Vietnam was left with the severe environmental damage brought on by the heavy use of fertilizers and pesticides to grow a crop in an inappropriate environment, as well as the social hardship caused by declining crop values. Farmers around the country faced poverty and ruin. Under pressure from farmers desperate to make ends meet and an impoverished population that needs wood for fuel and construction, Vietnam's remaining natural forests are being reduced at a rate of 250,000 to 500,000 acres annually. At the present rate, they will all be gone by the first part of the next decade.[10]

Yet the effect was not limited to Vietnam. Coffee farmers around the world were pulled into poverty by the falling prices that were triggered in part by Vietnam's overproduction. In desperation many attempted to increase production by expanding into land that was once pristine rainforest and escalating their use of chemicals. Others have abandoned their farms to livestock grazing. Some of the rarest animals in the world, such as tigers, elephants, rhinoceroses, gibbons and orang-utans are threatened by the increase in coffee production.[11] Research in one of the world's most important lowland tropical forests in southern Sumatra found that deforestation rates 'are directly related to the price of coffee paid to farmers; during peak prices, deforestation rates double as farmers and outside speculators clear additional forest in hopes of a quick profit.'[12]

Notwithstanding the considerable social and environmental consequences, IMF-mandated structural adjustment programs did not accomplish their main goal. The economic performance of countries that adopted structural adjustment programs was no better than the performance of those that did not undertake such reforms. The Fund's own research found that growth rates in countries that had adopted structural adjustments were no better than in those that had not.[13] A review of African cases demonstrated that manufacturing declined, exports shrunk, and capital accumulation slowed in countries that adopted IMF-sponsored programs.[14] In fact, mounting evidence indicates that by a variety of measures, these programs have failed.

Recently, with too many countries unable to meet their mounting debt, and protest rising, the World Bank and the IMF initiated a debt relief program. But there was no real shift away from the Bank's ongoing policies. The 1996 Heavily Indebted Poor Countries (HIPC) initiative offered minimal debt reduction for a few countries. Participation required the completion of a six-year program of structural adjustments. The aim of the Bank's efforts was for countries to achieve 'debt sustainability' – that is, to ensure that they can continue to make payments on their debt. Three years later, after rising public criticism, an 'enhanced' HIPC initiative was announced. Again, only open to a few countries, the initiative now required a three-year program of structural adjustments that could be extended into even more years if the Bank deemed it necessary. If countries' delay required reforms, say by continuing to support low-cost energy or food, 'overspending' on public health, or not privatizing key industries, they would not receive relief. For those few countries that did qualify, the effect was hardly adequate. For example, after debt relief under this program, Mali was still required to pay $88 million each year to service its debt. In a country where a full quarter of the children do not live to 5 years old and annual spending on health per person is about $5 a year, the payment on their debt was nearly twice what the government spent on healthcare.[15] Overall, neither IMF-imposed economic policies nor debt relief initiatives have substantially reduced debt burdens.[16]

Subsidizing destruction

Because no single institution plays a more powerful role in the impoverished world than the World Bank, it is particularly egregious that the Bank has all but ignored the global environmental crisis, disregarded the potentially cataclysmic consequences of forest loss, global warming, loss of biodiversity, water scarcity, and other pressing environmental issues, and continues strongly to promote environmentally destructive industries, unsustainable exploitation of natural resources, and a narrow fixation on fossil-fuel-based projects. World Bank projects were responsible for roughly 7 billion tonnes of CO_2 emissions annually in the 1990s, and the Bank's investment in fossil fuels is increasing.[17] More than three-quarters of the Bank's lending on energy projects is spent on climate-changing fuels; while funding on projects meant to avert climate change is relatively minute. Not surprisingly, a full 90 per cent of the Bank's fossil-fuel projects benefited multinational corporations in wealthy countries. Because World Bank funding serves as a catalyst for further public and private investment, each dollar spent by the Bank can be followed by about $5 of corporate investment. More generally, the Bank sets the terms and conditions for major international investment and development strategies. So these numbers are even more disturbing than they appear at first glance.

The Bank has given short shrift to alternative energy possibilities. For example, it has strongly encouraged a focus on coal to meet China's future energy needs, despite the fact that coal is the dirtiest, most carbon-intensive of fossil fuels and China is already the single largest user of coal. Over $1.3 billion has been spent on four massive coal power plants in China, which will eventually release more than 2 billion tonnes of carbon dioxide into the atmosphere. Nevertheless, the World Bank is working with the Chinese government to facilitate an increase in foreign investment in coalmining.[18] The Bank has been a major proponent of expanding coal power plants and mining in Russia, Ukraine, Pakistan, India, and many other countries.

The impact of such projects is not limited to the threat of global warming. Coal mining not only threatens land degradation and serious disruption of the local ecosystem; it can cause acidic seepage into rivers and lakes, toxic dusts, as well as carbon, sulfur, and nitrogen leakages and waste. Nevertheless, in the late 1990s, a major loan to India allowed for the expansion of two dozen open-pit coal mines. World Bank officials acknowledged that they were 'rolling the dice' when they agreed to the loan.[19] The result was devastating environmental destruction from open-pit mines that were later filled with water, creating contaminated, stagnant lakes where there had once been farmland. Thousands of impoverished peasants were displaced without adequate compensation. The Bank's own review of the project revealed that policies toward resettlement, indigenous peoples, environmental assessment, and supervision were inadequate.[20] Unfortunately it was the people and land of India that paid the price.

Examples are all too easy to come by. Currently, a nearly 2,000 mile pipeline that would endanger fragile wetlands and subtropical forests of both Bolivia and Brazil is being considered by the Bank.[21] Yet another project under consideration at the Bank would drill some 300 oil wells in southern Chad and run a 600 mile buried pipeline to the Atlantic Ocean. Human rights abuses along the pipeline route have already occurred when hired gunmen killed unarmed civilians opposed to the project. The Bank's own independent panel of inspectors has criticized the plan. Environmentalists and social activists in Cameroon have pointed out

3.1 / Number of large dams around the world

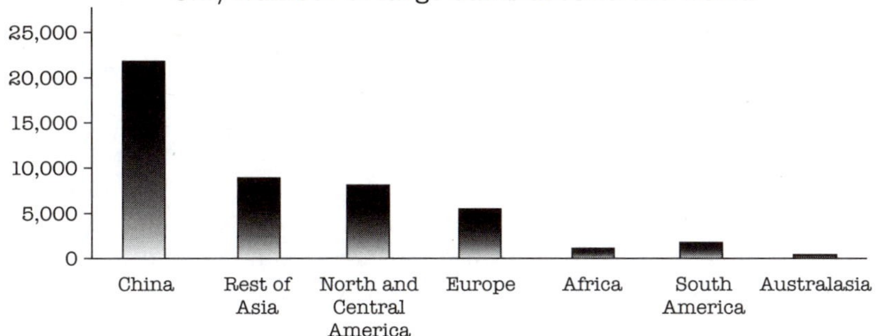

Source: *Dams and Development: A New Framework for Decision Making*, Report of the World Commission on Dams, United Nations Environment Program, Earthscan, London, 2000, p. 12. www.dams.org.

that the risks far outweigh any potential benefits. Still the World Bank is pushing ahead with its plan.[22]

The Bank directly funds many of the most environmentally egregious dams around the world. It has provided in excess of $50 billion for construction of more than 500 large dams in 92 countries. The environmental consequences of these dams include ruined wetlands, devastated fisheries, flooded forests and destroyed habitats of endangered species. Beyond their considerable environmental costs, these World Bank projects have forcibly displaced about 10 million people from their homes, ruined their livelihoods, and destroyed communities and cultures in the process.[23] Often, the violence and human rights abuses perpetrated in the process are considerable. At a minimum, as the Bank's own review found, the vast majority of people displaced by these projects faced landlessness, joblessness, homelessness, health risks, food insecurity, loss of access to common property assets, and social marginalization, while receiving little or no direct benefit from projects.[24]

Not only are the displaced populations pushed to the limits of poverty and starvation; they often represent an enormous burden to the natural resources and communities in the area of resettlement.[25] The overall impact can be crushing and is often most severely imposed on women, children, and indigenous peoples. A review by the World Commission on Dams, an independent body supported by the World Bank and the World Conservation Union, found that while dams can have considerable benefits for development, 'in too many cases an unacceptable and often unnecessary price has been paid to secure those benefits, especially in social and environmental terms, by people displaced, by communities downstream, by taxpayers and by the natural environment.' The report further found that 'lack of equity in the distribution of benefits has called into question the value of many dams in meeting water and energy development needs when compared with the alternatives.'[26] Yet the World Bank remains the largest single source of funding for large dam construction.

There are too many examples. In the late 1970s, the World Bank financed the Chixoy Dam in Guatemala; unfortunately 75,000 Maya Achì indigenous people were in the way. When the Achì resisted mass government resettlement, the government began an intimidation campaign by paramilitary death squads. With no consultation and little warning, villagers were forced from their homes. Those who opposed were labeled guerrillas and violently removed or killed. Roughly 400 were left dead. Survivors who fled to the countryside were hunted down by paramilitaries and killed. More recently, the World Bank funded the 42-mile long Yacyreta Dam that straddles the Argentine and Paraguayan sides of the Paraná river. The dam has displaced more than 5,000 people, with very few properly compensated for their losses. Argentine environmentalists warn that the dam may threaten 4,600 square miles of fragile wetlands that are home to endangered species, threatening not only the ecosystem but the area's economy.[27]

Such projects have inspired protests around the world. More than 3,000 of the more than 25,000 villagers adversely affected by the World Bank-funded Pak Mun Dam in northeast Thailand occupied the area next to the dam for seventeen months. They demand that the dam's gates be opened and that the river and their lives be restored. The World Commission on Dams has found that fifty-six species of fish in the river have disappeared since the dam was built, and fish catches have fallen by up to 80 per cent. Nevertheless, the World Bank has called the project's resettlement 'satisfactory' and rejected the demands of the local population for justice. To make matters worse, although the dam promised to generate 136 megawatts, it barely generates 40 megawatts in high-demand months, making it an economic failure. This is only one example of a growing international movement of dam-affected communities demanding that the World Bank accept responsibility for its failures.[28]

In South Africa, the World Bank is currently funding the troublesome Lesotho Highlands Water Project (LHWP) and facing intense popular protests. This is Africa's largest infrastructure project, involving five dams, miles of tunnels through local mountains, and a small hydropower component. The project's aim is to deliver water to South Africa's industries and metropolitan Johannesburg. Critics have pointed out that while the country's wealthiest citizens in the city's affluent suburbs will enjoy watering their lawns and filling their swimming pools, most of the poor in South

Africa's townships will be unable to afford the expensive water. Moreover the project has failed to compensate landowners for their loss, reneged on its promise to rebuild homes damaged by construction, and caused massive population displacements that have worsened social hardship, overcrowding, and food shortages in the region.[29]

World Bank-funded operations clearly exceed any reasonable level of moral acceptability. At least thirty Bank projects in twenty impoverished countries around the world involve waste incineration that generates dioxins, mercury, and other hazardous pollutants, even though safer and more economical alternatives are available.[30] Because dioxins are known carcinogens and linked to birth defects, and mercury is toxic to the central nervous system, kidneys and liver, such practices have been prohibited or phased out in the US and Europe.

Bank support enabled an American company to begin copper and gold mining in New Guinea, leading to over 120,000 tonnes of toxic mining waste being dumped into local rivers and killing over 30 square kilometers of lowland forest. In 1995, a gold mine in Guyana that had been made possible by World Bank support leaked 764 million gallons of cyanide-tainted waste into the main water source in the country. The river is still tainted and its fish contaminated. The Canadian company that ran the mine has rejected demands for adequate compensation from the 23,000 people living along the river.[31] The Bank also provides insurance for the Australian-owned Lihir gold mine in Papua New Guinea. Every hour of its operation this mine dumps 4,600 tonnes of toxic waste into the ocean through a subterranean pipeline 500 foot below the surface. It also dumps 20 million tonnes of contaminated soil and rock into the sea from barges each year.[32] The result is disastrous for sea life and the local populations that depend on the ocean for their food and livelihood. Like many operations backed by the Bank, such practices are now strictly prohibited in the US and in violation of international law.[33]

The World Bank is also a major force behind the continued destruction of the world's forests. If the Bank had any intention of curbing deforestation, it could easily do so in Brazil, a country that has a large share of the earth's remaining tropical rainforest and is one of the Bank's largest borrowers. However, Bank-imposed trade polices have instead led to expansion of the timber industry and industrial agriculture in Brazil, both of which are major causes of deforestation. Indeed, the Bank has allowed timber industries a free hand in depleting forest around the world. In one of many cases, the International Finance Corporation (IFC), the World Bank's private-sector arm, loaned millions of dollars to a European company to support the logging of 1,185,600 acres just outside the Ndoki national park in central Africa, a region believed to be the only forest in the world that may never have been inhabited by human beings. The region is home to forest elephants, lowland gorillas, chimpanzees, and other precious species. Not only was a huge area of forest outside the park razed, but, predictably, loggers secretly invaded the park to cut highly valued timber in protected forests. In too many cases throughout Asia, Africa, and Latin America, the Bank has not only directly funded logging in pristine forests; it has forced policy changes that favor foreign logging companies and promoted road-building that opened up new forests to logging industries.

A particularly disturbing case occurred in the 1980s when the World Bank provided over half a billion dollars to remove nearly 4 million people from Java and resettle them on surrounding islands in an effort to relieve population pressures on the densely populated island.[34] Over half of the area in the outer islands of Sumatra and New Guinea, where these communities were resettled, was virgin forest; and most of this land was already inhabited by indigenous peoples. While all so-called transmigrants were to be volunteers, relocations were not always voluntary and were often used by the government to subdue political dissidence. The result was not only the destruction of over 9 million acres of forest but also increased poverty for both those who were displaced and those who already lived in the resettled area. Left to attempt farming on forest soils that are nutrient-poor, and with a monsoon climate that renders the region highly vulnerable to erosion and runoff, migrants were beset by miserable yields, flooding, and soil depletion. Hundreds were killed in the ensuing violence between settlers and indigenous communities; many more were forced to flee. In some areas, up to half the settlers simply abandoned their sites.[35] Similar shortsightedness led to largely failed programs in Brazil, but only after the clearing of huge tracts of rainforest.

By the 1990s, outrage over the Bank's devastating schemes in the Amazon, Southeast Asia, and West Africa led it to adopt a ban on direct financing of logging activities in primary forests.[36] This seemed like a step forward. However, a lack of enforcement and lackluster implementation by Bank staff, who (an internal review found) often simply ignored the new guidelines, meant that little really changed.[37] More crucially, these new policies were not applied to structural adjustment programs, one of the driving forces of deforestation. Since structural adjustment loans represent more than half the Bank's annual lending – and the proportion seems to be growing – leaving this out of the picture makes any policy change hopelessly inadequate. The Bank's own internal study showed that only about a quarter of the Bank's forest projects were considered sustainable; and less than a third of the forest components of other projects were sustainable. Out of these, only about half had favorable outcomes. As its own review concluded, 'Bank influence on containing rates of deforestation in tropical moist forests has been negligible.'[38]

A new policy proposed in 2002 allows market forces to define deforestation policies. This program may well open the floodgates for large-scale industrial forestry investments and thereby accelerate global deforestation. Among other problems, only areas deemed to be 'critical forests' will be closed to further extraction; and which forests are to be defined as 'critical' will be decided by Bank officials without required input from local peoples. The result will certainly be increased Bank-funded logging in forest around the world, and the trashing of precious ecosystems and vital habitats.[39] The proposal has been widely criticized for weakening restrictions, enabling increased forest destruction, and perpetuating the marginalization of local communities in the development process.

However, the Bank's most powerful impact on the world's forest is not through direct funding for forestry projects but through the continued imposition of structural adjustments. Structural adjustments are closely tied to currency devaluation, export promotion, and removal of price controls; and all these policies increase exports of agriculture, logging, mining, and oil, and thus encourage deforestation.[40] In Latin America, Asia and Africa, forests are destroyed to clear lands for export industries such as industrial agriculture, beef production, and industrial shrimp production with Bank support. Even when the Bank does not directly support these industries, road building in forest regions, always a high priority on the Bank's agenda, opens up vast virgin forests to such destructive industries. Moreover, with public institutions and regulations weakened by the budget cuts required by the IMF, the violation of environmental legislation, reduced public spending on environmental protection, and weakened enforcement of conservation programs and environmental laws are widespread.

Once again, the link between social inequity and environmental destruction is clear. Land tenure systems that dispossess the majority of the population from their livelihood in favor of a few wealthy landowners, complete disregard for the rights of indigenous peoples and their lands, dire poverty that drives the poor to turn to the forest for cooking and heating fuel, and the ever-expanding and insidious search for exploitable natural resources by corporations, are the principal forces behind deforestation.[41] As long as the IMF and the World Bank continue to intensify, rather than challenge, these underlying drives, deforestation will continue.

The World Bank is just one example of a global set of institutions and practices that subsidize exploitation and environmental destruction for the sake of corporate profit. Export credit agencies (ECAs), government agencies with a mandate to promote overseas investment, have a far greater cumulative financial impact on investments around the world than does the Bank.[42] Virtually every wealthy country has some form of ECA, and the US has three. These agencies essentially use taxpayer money to make it cheaper and less risky for wealthy multinational corporations to do business overseas. But their aim is not to reduce poverty or promote development; rather, their sole purpose is generally to promote exports and foreign investment and therefore foster greater profits and growth for their country's industries.[43] So, while the Bank provided about $8 billion to projects with adverse environmental impacts in 2000 (a disturbing amount to be sure), such projects received over eight times that amount from ECAs.[44] These agencies operate away from the public eye, with no obligation to consider or release information on the potentially devastating social or environmental impact of any project that they fund.

Such export credits are clearly distributed with political intent. In the US, the ten top recipients of support accounted for more than 85 per cent of all such loans. They include companies like Boeing, General Electric, Raytheon, and Bechtel. As leading employers in many congressional districts and major campaign contributors, they enjoy the strong support of US policymakers and the bounty of corporate welfare. A considerable proportion of export credits go to support oil and gas development and energy-intensive industries. Arms sales, military hardware, and nuclear power take up the slack.

The World Bank has never been forced to bear the consequences of the destruction it has helped bring to communities and ecosystems around the world; and ECAs are even more careful to dump the risk onto someone else. Of course, both expect repayment even for failed or environmentally catastrophic projects. More than this, in order to attract much needed investment, the governments of impoverished countries are often required to provide 'sovereign counter guarantees' ensuring that they will assume the debt if the local industry fails to pay. That means the private debt of a corporation can frequently turn into an addition to the already large public debt of a country. Furthermore, providing the extremely favorable terms and conditions required to attract investment can lead to massive budgetary liabilities and increased debt. A full 40 per cent of the official debt of the world's impoverished countries is export credit related. If something goes wrong or a project seems unlikely to meet expectations, there is still every reason for the company to push forward, since the host government has assumed much of the risk.

Thus, even when the World Bank, facing public criticism, will not fund a project, export credit agencies, with the sole aim of facilitating the profits of Western industries, will. A clear example is the Three Gorges Dam project in China. One of China's most respected writers has called the project 'a symbol of uncontrolled development' and 'the most environmentally and socially destructive project in the world.'[45] The project will displace more than a million people and permanently flood thirteen cities and thousands of towns and villages.[46] Not only will tens of thousands of acres be flooded, but also the forced relocation of this massive population will intensify deforestation and soil erosion in surrounding regions. Endangered species such as the white-fin dolphin will suffer. Most crucially, many experts feel the dam will be unsafe and could potentially collapse after completion.[47] Yet this is only one of a seemingly unending number of projects around the globe that put a narrow fixation on profits and an unrestricted market above concerns for the fundamental security of humanity and nature.

Nevertheless, the Washington Consensus continues to dominate policymaking and define a narrow range of options for impoverished countries. An ideologically driven faith in the efficiency of an unregulated market has blinded policymakers to their own failures, or perhaps it has simply allowed them to justify a push for their own interest despite these failures. Either way, free trade has become a deceiving slogan used by the wealthy to ensure the continued exploitation of the poor.

Free trade isn't free

Regional free-trade agreements such as the North American Free Trade Agreement (NAFTA) have defined and enforced this same constricted focus on unregulated markets at the expense of public health, the environment, and the quality and security of people's lives. On the day NAFTA entered into force, a ragtag group of villagers in the poor Mexican state of Chiapas took up active resistance. Calling themselves the Zapatista National Liberation Army, in honor of the popular early-twentieth-century revolutionary Emiliano Zapata, they protested policies that put land in the hands of a few wealthy landowners who raised coffee or beef for export while denying peasants enough land to survive. Their fear was fed not only by the pain free-market reforms had already wreaked on their own lives, but by the devastating impact it had on many other areas of Mexico.

The squalor in the assembly plants located just across the American border, in the so-called maquiladoras of northern Mexico, is well known. Seeking to benefit from a desperate population that is willing to work for starvation wages, American manufacturers began setting up shop in northern Mexico in the 1960s and now number in the hundreds. In the first two years of NAFTA, their numbers jumped by more than 20 per cent.[48] Wages are low; the average worker is paid 75 cents an hour.[49] But that's not the only attraction; more than one quarter of the operators in the 1980s cited lax environmental enforcement as a reason for their decision to locate there.[50] The social, public health, and environmental crises that have resulted are dramatic. Growing communities of Mexican workers live in sprawling shantytowns with no clean drinking water, sewerage, or other basic services. Girls and young women, most between 14 and 20 years old, work grueling ten-hour days, six days a week. Work conditions are often brutal, Mexican labor codes are rarely enforced, and injuries are commonplace. By exploiting these desperate workers, industries are able to assemble imported components and ship the finished products back to the US at incredibly low cost. Despite the claims of free-trade advocates, things are not getting better; the minimum wage for workers on the border is the same as it was ten years ago when NAFTA was first signed, about $4.20 a day.[51]

Freed from the regulations they would face north of the border, these plants poison the air and dump their toxic waste into the Rio Grande river, from which 95 per cent of the region's population gets its drinking water. A National Toxics Campaign study in 1991 found methylene chloride (a suspected carcinogen) in canals adjacent to an industrial park at levels 215,000 times the US ambient water quality standard.[52] Intimidation and bribery help maquiladora industries take advantage of an already weak capacity to enforce the Mexican environmental regulations and avoid the cost of proper and legal disposal.[53] More and more companies fail to report their hazardous waste shipments.[54] Transportation spills, abandoned waste sites, and illegal dumping abound. One survey found 1,247 illegal dump sites and estimated that there were some 20,000 more.[55] Little wonder children in surrounding US and Mexican towns are born with serious birth defects.[56]

Some might argue that these environmental costs are a necessary trade-off for modernization; but economic conditions in Mexico are clearly not improving. Labor productivity in Mexico's manufacturing sector has increased since the inception of NAFTA, which means greater profit for industry; but at the same time the real value of wages has decreased. Growth in this sector has not kept up with job losses in areas hurt by imports, leading to an overall loss of jobs for Mexican workers.[57] At the same time, Mexican farmers have endured a massive increase in imported corn and a 45 per cent drop in the value of their crops. Farmers and rural workers have had to leave their homes in search of jobs in cities and in the US. Once again, women have endured the brunt of this transformation in Mexico's rural areas. A major report from the Carnegie Endowment for International Peace released on the tenth anniversary of NAFTA made the impact of the free-trade agreement clear. NAFTA, according to the study, had failed to generate jobs in Mexico, had hurt thousands of Mexican farmers, and had a 'minuscule' impact on jobs in the US. 'On balance,' a lead researcher noted, 'NAFTA's been tough for rural Mexicans.'[58]

Yet NAFTA means much more to corporations than free access to low-wage workers. Embedded in the NAFTA agreement is a system of rules that empower corporations to force changes in a country's social or environmental policies if they are not favorable to the continued profits of wealthy industries. The so-called Chapter 11 provision allows corporations to sue a local or national government if its policies harm the company's profits, for example. So, if a country were to pass legislation for environmental protection or public safety, a corporation could sue for the impact the environmental protection might have on its profits.

The scope of Chapter 11 is unprecedented. Even private investors, a minority shareholder in a company, or virtually anyone who has a stake in the profits of that company, can bring a suit against a government.[59] Any 'law, procedure, requirement or practice,' both legally binding and not, is open to challenge. Investors and companies can even demand compensation for the loss of potential future profits caused by the state's efforts to protect its people and environment!

Moreover, investors and corporations don't even need to make their case in the courts of the country concerned. An alternative adjudication system was put in place, allowing wealthy investors to sidestep the judicial process and its safeguards for the rights of peoples, and make their claim in a secretive system of arbitration that has the power to make rulings that cannot be appealed. A panel of three judges who have experience in trade and investment law, but not necessarily in social or environmental issues, decides each case. There is no public access to the proceedings; and little or no information on the case is provided to the public. The judges are not required to seek expertise if the case has environmental or social impact; nor are they required to hear briefings or opinions from outside experts. In short, a small group of hand-picked trade lawyers can do as they like with a country's social and environmental policies. The result is a clear end-run around the democratic process.

The result has been identified as a 'chilling effect' on the development and enforcement of social and environmental policies. To be acceptable under NAFTA, environmental legislation is required to meet a series of demanding tests: (1) it must be shown to be necessary to protect the environment in the country or region adopting the measure (so a country cannot enact a measure to end environmental destruction outside its borders); (2) the necessity of a trade measure in order to achieve this aim must be made clear (in terms that trade lawyers find convincing and 'scientifically' conclusive); (3) the measure must be shown to be the least trade-restrictive measure possible (an ambiguous requirement that seems to allow the tribunal to reject measures at will). These requirements, and the broad power given to corporate investors through Chapter 11 to enforce them, have made the NAFTA agreement a valuable weapon in the corporate assault on the environment of North America.

An early lesson on the importance of this weapon came in 1996, when the Virginia-based Ethyl Corporation used Chapter 11 to force Canada to reverse its ban on the toxic gasoline additive MMT. Because MMT causes irreparable damage to a car's pollution-control devices, and thus increases the emission of the greenhouse gas carbon dioxide, deadly carbon monoxide, and dangerous volatile organic compounds, Canada banned the import or inter-provincial sale of MMT. The additive is also a clear neurotoxin with unknown health risks. Alternative additives, including alcohol, can increase octane ratings with less environmental and health risks; so most Western countries and many states in the United States have chosen to ban the additive. However, faced with the threat of diminished profits, Ethyl sought to have the US initiate a state-to-state proceeding against the Canadian ban. When the US government refused, Ethyl continued to pursue the case alone. Despite the fact that there was no producer of MMT in Canada (Ethyl was the only North American manufacturer) the company used the lack of a domestic ban on the toxin to argue that Canada was unjustly impacting foreign producers in violation of NAFTA's requirement that domestic and foreign producers be treated equally. Ethyl demanded $250 million in loss of future profits, claiming that Canada's regulation was an illegal expropriation of the company's potential profits. Faced with the threat of a severe loss, in 1998 the Canadian government agreed to reverse its ban, pay Ethyl $13 million, and issue a public statement saying that MMT had no health or environmental impacts!

This is not an isolated example. A year later, a US state was on the losing side of a very similar case. In 1999, the Methanex Corporation of Canada sued the US for $970 million in response to a California ban on the gasoline additive methyl tertiary butyl ether (MTBE). Because MTBE allows gasoline to burn cleaner, it was added to fuels throughout the US. However, when its severe neurotoxic qualities and its probable carcinogenic effect became more clear, MTBE presented a patent environmental and public health hazard. MTBE had already contaminated at least thirty public water systems in California and proved virtually impossible to remove from groundwater. Faced with the prospects of lost profits, Methanex challenged the California ban under Chapter 11. Claiming that the ban unfairly limited the company's future profits and market share, Methanex demanded $970 million in compensation. Because the United States, not California, was the defendant, the state had no role in its own defense; even the California attorney general could not speak before the tribunal.[60]

The threat of Chapter 11 litigation has had a clear dampening effect on the willingness of local officials to fight for the protection of public health and the environment. With a potential Chapter 11 challenge looming over the heads of local and national leaders, making or enforcing social and environmental regulations entails the added economic and political risks of being held liable for the lost profits – or potential profits – a company or investor may claim would have been possible if that regulation had not been implemented. Many times, the mere threat of a Chapter 11 suit is enough to prevent new environmental policies from being enacted.[61] Corporations can simply harass state and local officials into avoiding or eviscerating social and environmental protections. The cost to corporations is low, a few thousand dollars to initiate a suit; and the payoff is high, a world free of pesky, profit-reducing regulations meant to ensure public health or protect the environment.

A clear example was provided when local officials closed down a hazardous waste disposal site in San Luis Potosí, Mexico, on the grounds that it was not environmentally sound. The facility had a long history of serious pollution problems, and community opposition to the site had been intense. After two years of negotiations, Mexico's federal government had granted the site a license, but local officials remained steadfast in their opposition. Unfortunately for the local Mexican authorities, the site was owned by a subsidiary of the US company Metalclad. The company sued under Chapter 11, claiming that by not granting it a license to operate its hazardous waste facility, local Mexican authorities have unduly impinged upon the company's profits. Local Mexican officials had no access to the proceedings, and the arbitration panel had no obligation to listen to or evaluate the findings or opinions of any environmental experts or community advocates. Not surprisingly, Metalclad won their claim, forcing the Mexican government to pay $17 million to Metalcad in punishment for their attempt to protect the environment.

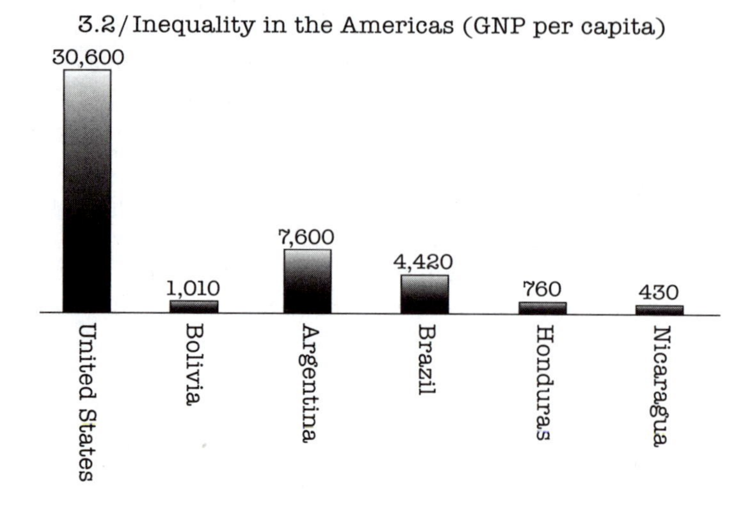

Source: World Bank, *World Development Report 2002: Building Institutions for Markets*, Oxford University Press, New York, 2002.

Clearly, when we define trade agreements to prioritize only corporate profits and wealth, at the exclusion of all other concerns, we set the rules for a very destructive process. But we may only be seeing the very tip of the iceberg. If this right to compensation for measures enacted to protect the environment is extended beyond NAFTA, it could mean hundreds of millions of dollars in compensation for foreign investors. If this practice is applied to domestic investors as well, any loss of any conceivable potential profit or value due to environmental policies would require compensation from taxpayers. The tab would quickly run into the hundreds of millions of dollars, making the enforcement of current environmental regulations very costly and the likelihood of future regulations very slim. The 'polluter pays principle,' which is well established in domestic and international law and requires the corporation that produces the pollution to pay for the cleanup, would be replaced with a policy that requires taxpayers to pay polluters for the loss of profits that result from rules that stop them polluting the environment. Essentially, polluters would own the right to pollute the environment with impunity and would need to be compensated if this right is infringed upon!

Things are likely to get worse. On the tenth anniversary of NAFTA, despite mounting evidence that the agreement has had a serious negative impact on both US and Mexican workers, the Bush administration signed a new Central American Free Trade Agreement. The last hurdle in the formation of the Central American agreement was cleared recently when the US House of Representatives narrowly approved the agreement in

3.3 / Poverty in the Americas

	Percentage below the national poverty line	Percentage living off less than $1 a day
United States	13	1
El Salvador	48	40
Haiti	65	-
Jamaica	34	5
Nicaragua	50	44
Brazil	29	17
Colombia	7	19
Ecuador	30	35
Venezuela	12	31

Source: *World Resources, 1998–99: A Guide to the Global Environment: Environmental Change and Human Health*, World Resources Institute, United Nations Environment Programme; United Nations Development Programme, and World Bank, Oxford University Press, New York, 1998.

a late-night session. It is a step toward the expansion of a NAFTA-like system to countries throughout South and Central America as the Free Trade Area of the Americas (FTAA). With the express decision to exclude environmental issues from the agenda, the FTAA would inflict NAFTA's destructive model of privatization, deregulation, and investor disputes throughout the continent, creating the world's largest free-trade zone and one of the most far-reaching trade agreements ever conceived. Expanding beyond NAFTA's scope, the FTAA will include services as well as goods. This would mean that public functions such as healthcare, education, or social services would be included; and cultural, social, or environmental polices could be more directly challenged.[62] The investor provision of NAFTA is likely to be included in the FTAA; so any state, corporation, or individual that can claim a stake in the case can challenge another country's laws or procedures. The FTAA may incorporate a binding dispute settlement mechanism that is much more powerful than current NAFTA procedures. This may well be the worst of NAFTA, expanded to include all areas of public life, and made far more stringent and mandatory.[63] Because the negotiations of the FTAA have been carefully kept behind closed doors, we cannot be sure what's developing. But a glimpse into this future may become possible as we examine the current impact of the WTO – on which many of the current developments seem to be modeled – and imagine it intensified.

The World Trade Organization

The power of the wealthy to redefine the world in their favor is nowhere more clear than in the emergence of the WTO. Unlike other international trade institutions, the WTO has the capacity to enforce its rules independently. It can require compliance from member countries, and it can impose sanctions if they do not obey. And the scope of the WTO's authority is extensive, including not just the trade in raw materials and all forms of goods, but also services, including everything from banking to water utilities. In fact, policies need not even be directly trade-related to come under the WTO's mandate. For example, the WTO requires dramatic changes in intellectual property rights policies, even for goods that are wholly produced, distributed, and consumed in a single country.

Although only member countries, and not individual investors, may launch challenges in the WTO, the Organization's review mechanism is much the same as NAFTA. However, because of the WTO's wider scope, worldwide membership, and independent enforcement capacity, the implications are even more severe. Failure to comply with WTO rulings can result in fines and punitive action. These sanctions are not optional; the rulings are binding on every member country. A government can avoid a WTO enforcement action only if every other country – including the country that originally launched the dispute – agrees to end action. A tribunal of trade lawyers nominated by the WTO secretariat makes decisions. The disputing parties can oppose a panel member only for 'compelling reasons.' Appeals are heard by the WTO Appellate Body, which is composed not of independent judges but of selected WTO employees.

Like NAFTA, the WTO's decisions are not restricted to commercial matters. Roughly one-quarter of WTO challenges have involved a public health or environmental policy. In each instance, the WTO ruled the policy to be an unacceptable barrier to trade, and countries were required to eliminate or weaken health, safety, or environmental protections to satisfy WTO requirements. Like NAFTA, the WTO requires that such policies run a gauntlet of vaguely worded tests before being deemed acceptable. WTO officials are free to make their own judgments as to the importance or viability of the environmental or public health issue in question. They are free to dismiss expert testimony that is divergent from their own view. They are free to balance these aims with their own goal of unrestricted commerce without seeking the advice of experts in the field. And, once they have made their decision, enforcement measures are unavoidable.

With its sights set narrowly on an unregulated global marketplace, the WTO has cast all other priorities out of the picture. Governments are forbidden under the 'national treatment' requirement to treat foreign services and products any differently from local, indigenous products and resources. No requirement of domestic content may be established. So corporations cannot be required to involve local community members in the design, ownership, or management of their endeavors. Governments are required to consider only 'commercial considerations' when making their own purchases. They may not use government purchasing policies to promote environmentally friendly or socially beneficial products or industries.

A crucial requirement is that governments must ensure equal market access for 'like products.' This means distinctions cannot be made based on the production process or the social or on environmental impact of that process. Since a product's 'production processes and methods' (PPM) may not be used to discriminate, governments lose their capacity to establish policy on how a product is produced.[64] If a product is made in an environmentally damaging way, it must be granted the same market access as a similar product that is made through a more sustainable method. Since not protecting the environment is often cheaper, this gives a market advantage to environmentally destructive industries and ensures that governments may not intervene to reduce that advantage.

This policy was most famously identified in the conflict between Mexico and the United States over tuna and dolphins. Over 200,000 dolphins were being drowned, crushed or otherwise killed each year as the 'by-catch' of the tuna industry. Because dolphins swim with tuna, they get caught in the huge driftnets that sweep everything in their way. With the US Marine Mammal Protection Act requiring the protection of dolphins, the US announced its intention to ban the import of tuna from countries that had not adopted dolphin-safe practices. Mexico appealed to the dispute settlement panel of the WTO's predecessor, the General Agreement on Tariffs and Trade (GATT), arguing that the US sanctions were based on processing standards and thus could not be imposed. In 1991, the GATT panel ruled against the US; and a few years later a similar ruling was issued in a comparable European challenge.

Thus Mexican tuna, however scandalous and destructive the mode of catching, must be given the same access to the US market as tuna caught in a more humane and environmentally sound way. The fact that the ban applied to both domestic and foreign tuna industries was held to be irrelevant. In the opinion of the panel, less trade restrictive measures were available, although what those measures might be was not made clear. Because the ruling predated the formation of the WTO, enforcement was not automatic. Indeed, because the US and Mexico were in the midst of negotiating NAFTA, and the dolphin issue had the potential to spark a popular backlash against free trade, Mexico backed off. However, later Mexico demanded that the initial ruling be enforced, leading the US to gut the Marine Mammal Protection Act over the objections of environmentalists and consumers.

At the same time, because trade policymakers can manipulate environmental policies to suit the interest of favored industries, policymakers from impoverished countries have developed a suspicion of restrictions based on production methods, seeing them as a disguised protectionism used to block products from the impoverished world or at least make them less competitive. They charge that 'green protectionism' is yet another excuse to exclude their products from Western markets; and the argument is not without merit. It's clear that environmental standards, even when they are not simply veiled trade barriers, are defined in terms that are most beneficial to the enacting country. For example, the US ban on tuna was only enacted after a domestic ban placed US tuna producers at a competitive disadvantage. US canners pushed hard for the legislation. This is not to say that the US law was unjustified, only that when environmental protections are crafted to suit commercial interests before environmental needs, they provide less than adequate environmental defense, are often inequitable in their application, and are vulnerable to legal challenge. This practice feeds a clear and well-deserved skepticism toward the veracity of environmental aims pushed by Western policymakers.

The central problem with such protectionism is that, even when serving real environmental concerns, it assumes the problems are in the impoverished world and the only solutions are defined by the world's wealthy.[65] In short, the poor are destroying the earth, and the wiser rich must step in to save the planet with unilateral action (conveniently crafted to their own advantage). In light of what we have set out so far in the book, the absurdity of this claim should be obvious.

This dilemma was evident in the WTO's very first case, when it struck down US Clean Air Regulations and required the US to amend its attempt to reduce air pollution or face sanctions totaling $150 million per year. In an effort to achieve the 1990 Clean Air Act's goal of a 15 per cent improvement in air quality over 1990 levels, the US Environmental Protection Agency required that cleaner-burning gasoline be sold in America's most heavily polluted cities. In an effort to achieve this aim in an effective and feasible way, the EPA decided to require gradual improvement based on the past performance of each fuel producer. For corporations that could produce accurate records of their past production, the required reductions would be based on actual past performance. If reliable information was not available, the manufacturer would have to produce gasoline that was no more contaminated than the industry average for 1990. Since it would be very difficult to obtain accurate and reliable information from all the foreign sources of gasoline, the EPA determined that they too would have to meet the 1990 industry average. Its important to note that these rules were in no way established to regulate gasoline trade; they were about renewing clean air in America's most polluted cities. And, in fact, the resulting 'Gasoline Rule' was a very moderate but positive step toward addressing a serious public health and environmental crisis.

Foreign refineries complained immediately; and in 1995 Brazil and Venezuela filed a formal complaint with the WTO. Even though they were only required to meet average cleanliness levels, they argued that since foreign producers were held to a different standard than some US producers, the new environmental regulation was inconsistent with free trade. The WTO trade panel, and later the WTO's Appellate Body, ruled that the US Clean Air Regulations were indeed in violation of the requirements set by the WTO; and, they added, the small provisions in the WTO agreement that supposedly allowed countries to enact regulations to protect their environment and natural resources did not apply. The US was required to rewrite its policies, adopting an industry-defined plan that the EPA had previously determined to be unenforceable and too costly.[66] In the end, foreign gasoline producers won the right to import gasoline into the US that was more polluting than the 1990 industry average. The men, women, children, rivers, lakes and forests in and around America's most polluted cities are paying a high price for free trade.

Any pretense at objectivity is preposterous. Panel members did not hesitate to second-guess environmental experts and policymakers. Indeed, they had no problem essentially overriding the assessments of democratically accountable decision-makers on the basis of ambiguous WTO requirements. Like all environmental or public health policies, the Clean Air Regulations had to survive a gauntlet of tests: it had to be the 'least trade restrictive' option; its 'primary aim' had to be conservation; all 'reasonably available alternatives' had to be ruled out. Each of these determinations is highly subjective and based in exceedingly vague conditions. Since the WTO panel members were inclined to use the most trade-friendly (and environmentally hostile) interpretation of each of these subjective criteria, it is hard to imagine any environmental regulation that could survive their scrutiny. In fact, none has.[67]

Moreover, WTO requirements can make it more difficult for countries to abide by international agreements and accords meant to protect the earth's environment and its species. Consider an important case from

1998, when the US implemented a policy designed to protect sea turtles as part of its commitment to an international treaty on the protection of endangered animals called the Convention on Trade in Endangered Species (CITES). The US banned the import of shrimp from countries that did not require the use of Turtle Exclusion Devices (TEDs) to reduce the harm to endangered sea turtles that become caught in shrimp nets. More sea turtles were killed by the shrimp industry than all other human activities combined, totaling tens of thousands of death each year.[68] The use of TEDs – essentially an escape hatch in the net – could reduce that number substantially while having only a small impact on shrimp yields.

In response to the US policy, India, Malaysia, Pakistan, and Thailand filed a formal complaint in the WTO. Not surprisingly, the WTO panel determined that the US ban was in violation of WTO requirements. The backlash to the turtle decision was serious. Sea turtles became a favorite cause of environmentalists' attacks on the WTO. Under growing pressure from environmentalists, the WTO ruled in 2001 that the recently modified application of the law by the US meant that it was more equally applied and made the process more transparent, thus meeting WTO objections.[69] Essentially, rather than countries being required to adopt national sea turtle conservation policies in order to export to the US, individual shrimpers who wished to export to the US were required to use TEDs. This left more turtles exposed to the risk of shrimpers; but it did not entirely negate the US conservation effort. While the outcome was not entirely unfavorable, the precedent that was set was not particularly positive. First, the final WTO decision upheld the restriction on trade barriers that are based on the method of production. So, any process-related restriction, including bans on goods made with child labor or in violation of basic human rights for example, was still subject to challenge in the WTO. And, since the appellate body based its decision on the US obligation under the existing international agreement, it is probable that in the absence of a pre-existing international accord, the US policy would not have been allowed. Thus a country may not be able to take independent action to prevent environmental or social harm.

There was more at play here than may be apparent. The objection from the countries that contested the US act was not based on the environment or sea turtles; rather, they objected in large part to US arrogance. The Asian plaintiffs, and policymakers around the impoverished world, argued that international restrictions should be based on international negotiations among all concerned nations, not unilateral action by one, powerful country. For many, the central issue was not the protection of endangered species, but the US dominance of the rules of international trade.[70] Once again, crafting environmental regulations to suit corporate interests made them less than adequate protections and vulnerable to challenge. And, once again, environmental requirements appeared to those in impoverished countries as protectionism dressed in green clothing, requiring practices, technologies, or new chemicals that they simply couldn't afford. While fishers in wealthy countries may be able to incorporate TEDs relatively easily, in the impoverished world such a requirement could present hardship for small fishers. US policy made no credible effort to address this problem.

It is clear that the character and application of international rules and requirements – even those that are nominally focused on environmental protection – favor the most powerful countries and their industries. The US does not *expect* requirements to be equally applied across the board. Although in the tuna and sea turtle cases, the US was an advocate of environmental labeling and restrictions based on production processes, it has been a powerful opponent of even voluntary labeling on other products. Pushed by industrial agriculture and biotechnology companies, the US has strongly opposed European effort to label genetically modified foods, for example. Called eco-labeling, identifying products according to their environmental impact allows consumers to choose among different products using social or environmental criteria. The now-eviscerated 'dolphin safe' label is a well-known example. In 1996, a corporate front group representing timber, plastics, electronics, chemical, and packing industries, perversely calling itself the 'Coalition for Truth in Environmental Marketing Information,' joined with large food manufacturers to ensure that the US blocked European eco-labeling plans. Industry lobbyists defined every aspect of the US position; and US representatives have dutifully pushed the corporate-defined program against eco-labeling at the WTO.

Similarly, the US has successfully argued that a European ban on hormone-treated beef is not acceptable because it is based on the production process. Since the US position on sea turtles required making exactly this same distinction, clearly what is good for the goose is not good for the gander. It's obvious that industry interests, not a coherent national policy, are driving US decisions. If an attempt to protect citizens from a potential health risk is rejected as 'WTO incompatible,' and merely informing the customer of the environmental impact or possible health risk of a product is deemed a barrier to free trade, then it is increasingly hard to imagine *any* environmental or health policy that could not be attacked as a barrier to trade.

All international agreements meant to protect the earth, called Multilateral Environmental Agreements (MEAs), may be threatened by the primacy of the WTO agenda. In the mid-1990s, when a hard-won international agreement on biodiversity was trumped by a corporate-defined intellectual property agreement, this was abundantly clear.[71] In 1992, a series of international meetings on biodiversity achieved the agreement of more than a hundred countries on the conservation and sustainable use of biodiversity and the equitable sharing of benefits from that use. The Convention on Biological Diversity (CBD) specifically put access to genetic resources under the control of national governments, and allowed those governments to regulate and charge outsiders for access to these resources. The idea was to make sure that wealthy industries would have to negotiate with impoverished countries to gain access to plants and indigenous knowledge within their borders. Further, agreement called on parties to respect and maintain indigenous and local communities and support and promote their stake in local biological resources. Although the US worked to weaken the agreement, after much contentious negotiation, it signed. The US has not ratified the agreement, but by the end of the decade 176 countries had both signed and ratified it. While officially this treaty had the power of international law, without US support the effect has been clearly marginal.

This was not an accident. The US was intent on the protection of intellectual property; and a meaningful agreement on biodiversity could get in the way. Since US industries were increasingly focused on services, and the actual production of many products was increasingly taking place in impoverished countries, industry leaders had a great need to ensure their continued profits by controlling their ownership of ideas, images, or brand identities. US companies needed to make certain that even if a product is made in an impoverished country and sold in that country, US industries would be assured of a share in the profits. To no one's surprise, ensuring the intellectual property rights that would protect these profits became the major aim of the WTO. Major corporations and their governments spearheaded an effort that culminated in the Trade Related Aspects of Intellectual Property Rights Agreement (TRIPS). The Agreement essentially requires all countries to adopt the US system of property right protection or risk WTO sanctions. With devastating implications, which will be discussed more fully in the following chapter, countries were required to ensure the protection of ownership rights to things that had previously and traditionally not been considered under the scope of 'ownership'.

So, essentially, the US made sure that an international agreement designed to protect the world's poor and indigenous populations' rights to their traditional knowledge and resources was undercut. Simultaneously, the US pushed to ensure that an international system of intellectual property law guaranteed the rights of wealthy corporations to their ideas and inventions. The kicker was that multinational corporations could now more easily take the traditional knowledge of a native tribe (for example, an ancient medicine or traditional spice) patent it in the US, and ensure their monopoly control of the profits. The indigenous tribe would be left in the cold, without the resources or knowledge to manage the complicated patenting requirements, and set to lose all rights to processes or resources that they had used for generations. As a result, local remedies, traditional practices, the knowledge of indigenous peoples, and other valuable and sometimes sacred resources and ideas are commonly usurped by Western industries. So-called 'bio-prospecting' by pharmaceuticals and agricultural industries threatens the wholesale piracy of genetic resources from impoverished countries. (We'll see plenty of examples later.)

The hypocrisy is evident. After all, the central legitimating claim of the WTO is that free trade solves all problems. But the fundamental character of the TRIPS agreement seems opposed to the very idea of an unregulated market. Every other aspect of the WTO seeks to limit the capacity of governments to control the market, say for environmental or social concerns. This intellectual property agreement, on the other hand, requires governments to regulate sharply and control the marketplace to ensure that Western industries get their cut. It's pretty clear that free trade isn't the priority of the WTO; maximizing profits for Western industries is.

Intellectual property rules clearly function to benefit western industries. More than 97 per cent of patents are held in wealthy countries. Under TRIPS, the normal burden of proof is reversed. Those accused of a patent violation must provide proof that they did not steal the product. The accused is, in essence, guilty unless they can prove their innocence. And because Western industries have the resources, connections, and assets needed to establish and defend a patent, they can seek out ideas, practices, and resources in impoverished communities around the world, patent them, and thus possess the sole right to control their use. This is not only clearly unjust, it's environmentally threatening. For example, the diversity of cultivated crops is likely to suffer because, since resources can be appropriated at will and without compensation, there's little incentive to conserve crop varieties and indigenous plants.

The WTO's virtually unrestricted scope means that the variety of practices and policies that could be ruled unacceptable by the WTO is extensive. Yet another example was provided when European countries announced that they would ban fur from countries that did not prohibit the use of steel jaw leg-hold traps. Long considered one of the most brutal forms of trapping, these traps seize and tear the animal's leg, face, abdomen, tail or other body part. In the pain and confusion, the animal can mutilate itself, tear joints, and break teeth in a desperate and futile attempt to break free. Severe and painful lacerations, broken bones, dislocations, and gangrene are not uncommon, as trapped animals die agonizing deaths. Nevertheless, the US, Canada, and Russian representatives complained that the European ban on these traps violated the free-trade requirements of the WTO, since the way the fur is produced constituted 'production processes and methods.' In addition, since most of the animals in question were not native to Europe, the rules had a disproportionate impact on foreign producers, they argued. Faced with US intimidation in the WTO, the Europeans backed down and settled on a vaguely worded and difficult-to-enforce multi-year phase-out of leg-hold traps that allowed continued unrestricted imports. Even if one is not offended by the cruelty of leg-hold traps, it's hard to argue that a country does not have the right to ban the sale of a product that is produced in a way that the population finds abhorrent.

It seems that even the most sophisticated environmental policies cannot survive WTO scrutiny. In an effort to address the critical threat from the electronic industry's toxic waste, the European Union proposed policies to minimize waste and shift the cost of environmental clean-up to the producer away from the taxpayer. Under the proposed European program, electronics manufacturers would have to provide for the proper disposal or recycling of their product after it is used. Electronic products containing lead, mercury, cadmium and other dangerous and contaminating chemicals would be phased out. And 5 per cent of the content of plastic electronic components would have to be recycled material – a paltry sum to be sure, but a start.

The American Electronics Association, an industry group representing major manufacturers such as IBM, Motorola, and Microsoft, immediately attacked the proposal. Industry representatives argued that the measures restricted free trade in the heavy metals used to make electronics; they claimed that the policies were more 'trade restrictive' than necessary; and they even challenged the scientific basis of restricting the use of the highly toxic chemicals. The latter point was particularly striking given the well-known threat these substances pose to human health and the environment. Nevertheless, the burden of proof was on the European Union to provide conclusive evidence of their danger sufficient to convince a panel of trade lawyers. At the behest of corporate interests, the US argued that these requirements would place too great a burden on the electronics industry and pushed for a voluntary and weaker program. Faced with pressure from industry and US representatives, the European Union backed away from its original proposal, reversing its ban on hazardous chemicals and eliminating the crucial recycling and disposal provisions.

Policies aimed at equitable development for impoverished countries are also threatened. Chiquita, the industrial agriculture giant, objected to a European policy that allowed preferential treatment for small banana producers in former European colonies in the Caribbean. The European policy supported local farms, which generally produce less environmental damage and are more beneficial to the local community. Indeed, the economies of many Caribbean islands depend solely on the banana trade this policy supported. With Chiquita controlling roughly half the European banana market, and Caribbean island producers maintaining only about 8 per cent of the market, preferential policies such as lower tariffs allowed these small farmers to continue to survive in a market dominated by a few industrial giants.

Although no bananas are produced in the US and very few Americans work in Chiquita's Central American facilities, the US challenged the policy at the behest of Chiquita. In 1997, a WTO panel decided in favor of the US challenge. A major contributor to both Republican and Democratic Parties, Chiquita could count on the US to dutifully represent its interests, despite the fact that the result would clearly work against broader US interests in the region. As EU Trade Commissioner Sir Leon Brittan pointed out, 'I really do not see why it is in the interest of the United States that poor countries in the Caribbean and elsewhere, which are not able to do anything other than grow bananas, should be driven into more dangerous economic activity such as drug trafficking.'[72] Nevertheless, facing more than $100 million in retaliatory sanctions, Europeans has no option but to change their policy. The impact on Caribbean independent banana farmers has been devastating. The number of Dominican banana farmers dropped by more than a third; and since for every banana farmer there are several others dependent on that farmer, the resulting social hardship has been widely felt.[73] While their bananas are produced with better labor practices and less environmental damage, in the absence of favorable trade rules small banana farmers throughout the Caribbean will have trouble competing with Dole and Chiquita's cheaper product.

The result is undemocratic and unjust. WTO policies have had a clear chilling effect on the adoption and enforcement of environmental protections around the world. The very character of the WTO process is skewed in favor of corporate interests and unrestricted commerce, placing policymakers in a defensive position. Since impoverished countries have less money, expertise, and influence to defend themselves before the WTO, their efforts to defend public health and the environment are particularly vulnerable to challenge. With all business taking place behind closed doors, the lack of transparency is disturbing and makes it impossible for social or environmental advocates, not to mention the general public, to evaluate the decisions, much less influence the process. Merely the threat of a WTO challenge can lead a country to shift its policy. The result is an unprecedented amount of power in the hands of corporations, corporate-friendly bureaucrats, and unaccountable policymakers.

The overall picture is all too clear. The current inequitable and unsustainable international order is imposed by a phalanx of international institutions and agreements that require and enforce corporate-friendly policies. These institutions serve as the rule makers, judges, and enforcers of a system narrowly fixated on the sacrosanct free market. Environmental and social concerns aren't simply secondary; despite some recent lip service, they seem to have no standing at all. The WTO makes no exceptions for a single international environmental agreement or treaty. Decision-makers in the WTO, NAFTA, IMF, or World Bank are neither required nor inclined to hear from environmental experts or advocates. All major decisions are made behind closed doors. All seem to share a common focus on complete capitulation to the interest of the world's wealthy corporations. These organizations are growing rapidly. NAFTA is expanding to include all of Latin America in a single agreement; and the WTO's authority and range is growing.

These rules make it ever more difficult for countries to protect their environment and population. The least common denominator in international standards is presumed as the WTO acceptable standard. If a country wants domestic standards on health, safety, or the environment that are higher than these bare minimums, it must survive WTO scrutiny. If a country wishes to restrict products that clearly harm the environment, it must pass a gauntlet of tests. There is, however, no floor on these standards, so no country need maintain any minimum health or environmental standard. The result is a powerful downward pressure on public health, safety, and environmental standards and restrictions. Meanwhile local men, women, and children, along with rivers, lakes, mountains and streams, are plowed under in the all-encompassing quest for greater wealth by the few.

As we will see in the next chapter, this pressure will surely intensify a process that is already in full swing. With unrestricted access to markets around the world, corporations can move to take advantage of cheap labor and lax standards. Industries seeking to avoid high labor costs or stringent environmental protection will relocate their manufacturing to countries that do not have or cannot enforce strong labor or environmental regulations. Desperate to attract investment, impoverished countries are pressured to compete for manufacturing in part by providing labor and environmental policies that are favorable to corporate interests. The weaker the protections and the lower the wages, the more desirable the country is to Western industries. The end result is a global 'race to the dirty bottom.' The revered free market will triumph; the earth and the majority of its people will suffer.

There's got to be a limit/4

Simple common sense tells us that we can only keep up this massive rate of consumption, waste and exploitation for so long. Something has got to give. We are on a runaway train headed for a cliff. We know we're going way too fast. If we cared to look, we could easily see the mangled tracks ahead, bent and splayed over the cliff's edge. Our current course is simply unsustainable. It's just hard to believe we're going to be stupid enough to keep going. Someone's going to stop the train, right?

Unfortunately, no one seems to be moving toward the brake. In fact, our leaders are shoveling more coal into the furnace, and most of the people on the train are shouting 'Faster! Faster!' We're fixated on continued consumption and waste, which we happily call 'growth.' We reject the very idea of limits, enjoying the landscape whizzing by out the window but refusing to look ahead. We place our entire faith in economic theories that cast growth as the unquestioned be-all and end-all of our existence and hail the free market as the benevolent and omnipotent distributor of all things. 'With economic growth, everything is possible,' we recite; 'without it, everything is lost.' We scream through the countryside, barely holding to the tracks, sending bolts and nuts flying, careening toward the chasm. From time to time, someone on the train tries to convince us that we're in trouble. 'No worries,' our leaders reply, 'our economic theories tell us everything is fine, the invisible hand will save us.'

In short, we're out of control. We have surrendered our fate to the market; and the market completely ignores the physical realities that threaten our existence. Our rate of consumption and waste is simply unsustainable. It may make us feel better to assert some economic theories that assure us continued expansion and consumption is possible and desirable. But these theories will not save us from the clear and evident reality: we are exhausting the world's capacity to support human life. Even if you are willing to take the morally bereft view that the suffering of billions around the world is not your concern, even if you're willing to ignore the gutting of our democratic institutions, the plain fact is that our current course is simply physically unsustainable. The commodification of such precious gifts of nature as food, water, and even life itself make the intolerable course evident.

Money is everything

We've built a global system based on the assumption that economic growth is our fundamental goal. Politicians promise greater and greater expansion, and people, deeply conditioned to want more and more stuff and ignorant of the real costs of their consumption, cheer them on. Corporations, with their eyes on ever-greater profits, tout theories that don't simply make growth an option, they cast it as a requirement of survival; and, they assure us, unregulated and unrestricted markets for literally everything and the unrestrained, self-centered pursuit of more will make us all better off. If this notion wasn't the core of our political and economic system, it would seem absolutely ludicrous.

Most often called neoclassical economics, these ideas are based in a resuscitation of the nineteenth-century economic thought that justified imperialism, child labor, and an unrestrained exploitation of natural resources. These theories repeatedly rehearse an unending faith in markets as the best way to distribute anything from widgets to education. Efforts to protect public health, the environment, or social welfare only get in the way of the 'invisible hand' of the free market, they argue. This infallible hand, we're assured, will make sure that everyone's better off, as long as we each just look out for our own immediate self-interest. This fervent, almost religious, support for the self-centered pursuit of more and the claim that growth is an unquestioned good and fundamental necessity of life is appealing to the 'haves,' but less so to the 'have nots.'

However, despite the assertions of corporate elites, their allies in governments, and some economists that this theory is dispassionate and scientific, it is clearly more ideology than theory. Its advocates pretend to describe inalterable rules and functions of human relations. The fact is this neoclassical economic ideology was invented and promoted by wealthy elites to allow some way of justifying a system that gave them control and legitimated their merciless pursuit of ever greater wealth.[1] Over the course of centuries, such theories have become imbedded in our matter-of-fact understanding of the world. As a result, they allow the destructive and exploitative global system to appear to be not just justified, but natural. It is neither.

This isn't the only possible economic model. Many economists recognize the weaknesses of neoclassical economic theories. Some have proposed an alternative model that recognizes natural resources and environmental degradation as integral elements in the economic process. They have been, at best, marginalized in the field, and their ideas have been relegated to the fringe of political acceptability. They're marginalized not because they're wrong – as we've seen, increasing scientific evidence indicates that they are right – but because these alternative philosophies would not serve the basic function of the neoclassical alternative: providing legitimacy to a way of doing business that would otherwise be not only obviously unsustainable but also immoral.

This is not to deny the value of markets in distributing certain things. I am simply claiming that markets are not the best way of distributing all things at all times. And, I might add, while the market is sometimes a useful tool in helping us achieve some goals, it is a damn poor way of deciding what our goals should be. The trouble is, increasingly, the only goals that matter are those defined by the market. Concerns over the health of the global ecosystem, justice, traditions, sacred beliefs, shared community, care and concern for fellow beings, are all left by the wayside. When deciding, in short, what our objectives are going to be, what price we're willing to pay to achieve that objective, what we're going to save and what we're going to sacrifice, everything is reduced to commodities bought and sold in a global marketplace.

As a result, the decisions we make are incredibly irrational even in strictly economic terms. A functioning economy requires people; and people require an environment that can support human life. Clean air, uncontaminated water, a stable climate, a living ocean, and a host of other crucial ecological factors are required for a functioning economy. Even by their own standards, simple economic calculations will tell you that the 'services' provided to commerce by a healthy ecosystem are worth trillions of dollars each year.[2] Yet, fixated on short-term profits, and with a narrowly constrained vision that excludes anything that cannot be immediately and directly translated into dollars, we continue to build a system that is fundamentally at odds with nature.

A recent essay contest organized by *The Economist* magazine and Shell Oil company reflects this ludicrous perspective. These respected institutions offered $20,000 for the best answer to this question: 'Do we need nature?' At best, the question is insanely arrogant. Who is the 'we' in their question? If it is multinational corporations, they have evidently decided that the answer is no. If it is humanity, the question simply doesn't make sense. We are nature. The imagined world in which this question makes sense, a world where human activities are somehow separate from nature, where nature is cast only as an unending reservoir of raw material to feed our ever growing consumption, is worse than erroneous, it is destructive. A better question might be, 'Does nature need us?'

This fixation on material growth narrows our capacity to define our goals and measure our progress in a meaningful way. A clear and important example is our use of gross national product (GNP) to measure the strength and success of nations. GNP measures the total value of all economic transactions in a country; and the rules of neoclassical economic ideology say that continued expansion of GNP is necessary to ensure a healthy economy. Healthy for whom? Not for the planet. Not for the billions who suffer daily to feed this perpetually growing consumption. Still, policymakers and corporate leaders continue to demand ever greater growth in GNP, and thus ever more consumption and ever more waste.

Of course, there's nothing wrong with 'growth' per se, if by growth we mean the expansion of knowledge, human welfare, understanding, and other good things. But this is not how growth is measured. GNP only measures a small sliver of what we say is important, and it even measures that poorly.[3] GNP doesn't measure human well-being or progress in any meaningful sense, it measures total commercial transactions. Every time a dollar changes hands, that's an addition to the GNP. The purpose of the transaction doesn't matter. A thousand dollars' worth of medical care for toxic poisoning is tallied the same as a thousand dollars spent on a child's education; but clearly these do not have the same real value and they do not reflect the same amount of *growth*. The effect is perverse: a company that pumps massive industrial waste into the local river will spark a greater boost to GNP – through the resulting costs of the cleanup effort, lawsuits, hospital bills, and funeral expenses – than a company that minimizes its waste. So, oil spills, lethal waste, or toxic poisoning show up as positive contributions in our interminable pursuit of more.

Beyond this insane accounting, our perpetual pursuit of more encounters another problem: quite simply, there is a limit to the amount of energy and resources the earth can provide. An extensive study at the Massachusetts Institute of Technology found that if the stream of resources consumed and pollutants generated by the human economy continues unabated, 'the most probable result will be a sudden and uncontrollable decline in both population and industrial capacity.'[4] As mentioned in the first chapter, we are already consuming 20 per cent more than the earth can sustainably offer. A continued fixation on growth isn't just a bad idea; it's suicide with a dull knife.

The most pressing limit may be the earth's capacity to absorb waste. Corporate elites and free-market economists, in an effort to assure us that there is no limit to our growing consumption, tell us that there is enough fossil fuel remaining in the earth for many more centuries of uninterrupted expansion. Well, let's look at what would happen if we were to use all that coal and oil. As the geologist and climate change researcher Jeremy Leggett has it, before we started burning fossil fuels there were about 580 billion tonnes of carbon in the atmosphere. There's now more than 750 billion tonnes of carbon floating around us, with 6 billion more added each year. Scientists warn that the addition of even another 200 billion tonnes would risk ecological catastrophe. Beyond 300 billion tonnes and the risk of global environmental devastation would be overwhelming.[5] Here's the rub: the total amount of gas, coal, and oil that is estimated to be left in the earth is about 10,000 billion tonnes. That's about fifty times the amount of carbon we can release into the atmosphere without inalterable and unbearable environmental consequences!

These limits are functions of natural laws; they are not open to negotiation. As the respected biologist Edward O. Wilson has pointed out, the earth's capacity to support continued growth of production and consumption is finite; the planet is 'exhausted and in trouble.'[6] Given our current rate of consumption, at even a modest 3 per cent rate of growth, we will devour more than six times the earth's sustainable capacity by 2050. The conclusion is inescapable: our fixation on continued economic expansion is simply not physically sustainable.

The inefficiency of markets

The invisible hand is choking the earth. The so-called invisible hand of unregulated and uncontrolled markets takes only immediate material factors into account. This ignores human and environmental consequences, and thus often rewards the most environmentally devastating practices. For example, about 75 per cent of the world's iron ore, and 95 per cent of the ore in the US, is extracted through open-cast mining. Vast pits are dug into the earth, or entire mountaintops are removed, to expose the desired ore. The destruction is devastating. The amount of waste is huge. The runoff can choke rivers. The released toxins can poison the countryside far beyond the extraction site. For corporations, none of this matters; the environmental costs do not touch the bottom line. So, the most devastating form of mining is also the cheapest and thus the most practiced.

The examples are plentiful. Over the past several decades, for instance, natural materials such as jute, an important natural fiber used in burlap and carpet, have lost ground to synthetic substitutes.[7] The production of the synthetic substitute for jute produces about 285 pounds of air pollution, including sulfur dioxide, nitrogen oxides, carbon monoxide and volatile organic compounds, for each tonne manufactured. The production also emits toxic pollutants such as ammonia, benzene, lead, and toluene. Although growing jute emits none of these pollutants, this gives it no advantage in the market. The synthetic substitute takes about six times the energy to produce as jute.[8] And, unlike jute, synthetics are not biodegradable and have very limited recycling ability. Nevertheless, since none of this figures into the bottom line, synthetics remain cheaper than jute and are replacing it at an increasing rate, driving jute farmers and the communities they support deeper into poverty. Jute is largely produced in impoverished countries, while synthetics are produced in industrialized, Western countries. Hence the decline in the world jute market has been particularly hard on farming communities that depend on the fiber for their livelihood.

Economists call it a 'market failure' when the free market does not lead to the best possible use of resources. Unfortunately, such failures aren't rare, they're common and integral results of the rules that define the system. When we rely on a system of valuation and decision-making that focuses only on material profit, we can expect outcomes that privilege material profit at the expense of social justice and our ecosystem. In economic lingo, factors that are excluded from market pricing are called 'externalities.' Things like the health of the environment or the welfare of our children are often excluded from the pricing mechanisms of the market, and thus virtually left without value. The presumed 'efficiency' of the market is simply a capacity to ignore the real cost of production.

With environmental damage excluded from the calculation of profit and cost, inefficiency is rife. The examples are endless. Car transportation requires the movement of about 2,000 pounds of steel, plastic, and rubber to transport a 150-pound person. That's less than 1 per cent efficient. Only 5 to 8 per cent of the energy used by a standard light bulb actually produces light. Carpeting remains in your home or office for about twelve years at most; it then stays in a landfill for as long as 20,000 years. That's less than 0.06 per cent efficiency.[9] To get just one pound of prawns, a trawler catches 10–20 pounds of marine animals that are then simply discarded, dead or dying.[10] That's 5 per cent efficient. Only 0.03 per cent of pesticides for aphids on beans reaches the targeted insect; more remarkably, just 0.0000018 per cent of DDT reaches the pest.[11] The rest simply contaminates the soil, water, and air, increasing the untabulated external costs. I could go on for pages, listing more and more disturbing examples of inefficiency. The fact is, market prices do a poor job of reflecting the real costs we pay for our consumption, and so they do a poor job of ensuring efficiency in our use of precious resources.

Pricing the priceless

When environmental costs do make it into the economic calculations, the effect can be even more perverse. The value of rich and precious forests and clean rivers, the purity of the air we breathe and the water we drink, are reduced to a price tag. This kind of logic underlies the practice of 'cost–benefit analysis' or 'risk assessment' that is now broadly used to determine policy choices. Its advocates argue that cost–benefit analysis is the most 'efficient' way of making decisions. In such analysis, everything is assigned a monetary value and hoisted on the scales of judgment. Whichever side has the greatest monetary value gets a favorable outcome. All relevant concerns, issues, and considerations are translated into a dollar value.[12] If you're trying to decide whether to reduce your use of toxic materials in a production process, for example, cost–benefit analysis would compare the costs of switching to alternative production methods (the training cost, materials, perhaps retooling) to the costs of painful, disfiguring, and debilitating diseases for your employees and the surrounding community. Perhaps the cost of a long-term public health crisis from the contamination of local water sources that are used by surrounding schools and homes will be factored into the equation. Of course, these won't be added to the balance as moral or ethical issues, but as potential monetary costs. All too often, it will boil down to one question: how many victims are likely to sue and how much are they likely to get? The destruction of the local countryside may or may not be factored in; but if it is, again, it will be in monetary terms. How much revenue might be lost if the forest and streams are decimated and the fish and fauna are dead? The damage to the environment beyond the immediate area is unlikely to fit in the equation at all.

Using similar analysis, so-called risk assessment can then be used to justify your desired option as an 'acceptable risk.' The question of acceptable to whom is never raised; instead the method is taken as an objective and scientific way of ensuring a practice is 'safe.' Yet there is no such thing as absolutely safe; such a finding is scientifically impossible.[13] Evaluations of relative safety and acceptable risk require, at their core, a value judgment. But, ultimately, the question that corporate and public decision-makers are inclined to ask is how much harm can be endured rather than how much must be avoided.[14] For example, the Food and Drug Administration consider a risk of one in a million as something of the gold standard in acceptable risk. Using such a standard, a contaminant that caused 250 deaths throughout the United States could be considered not only acceptable but having achieved the highest level of safety. The EPA uses an acceptable risk range that is as much as one hundred times lower. So, a target risk for carcinogens in drinking water could lead to 25,000 unnecessary deaths across the country.[15] For other contaminants, such as water-borne pathogens, EPA rules identify one infection in 10,000 as acceptable. hence a municipal water system could cause hundreds of infections and still be well within federal guidelines. Clearly, these determinations are not merely objective calculations. At their base is a fundamental belief about the value of human life.

The whole practice of reducing life, health, and the natural world to monetary values can never be dispassionate. Can the full value of a pond, river, or forest be reflected in the market value of the water or timber? Do we even have the right to assign such values to nature's precious resources? Can risk analysis address the full human scope of fears, values, and desires? This narrow fixation on a quantifiable price tag leads to the most bizarre conclusions. For example, in a study for the Czech Republic, a major consulting group concluded that encouraging smoking would be beneficial to the government because it would cause citizens to die earlier. This would in turn reduce government expenditures on pensions, housing, and healthcare.[16] In the perverse world of cost–benefit analysis, this makes sense!

In the end, all the fancy calculations and specialized lingo boil down to the personal priorities and preferred vision of the future held by the person doing the calculations.[17] Which is more valuable, a new highway that can cut ten minutes off your commute or a fragile wetland that is home to dozens of precious species? Clearly, how you answer depends on your personal perspective on the world. Practitioners of cost–benefit analysis often rely on surveys and other tools to assign a price tag to such issues and thus get around the charge of bias. However, simple surveys, measuring 'willingness to pay,' 'use value,' and even 'non-use value' are largely meaningless. Given such a survey, how much would you value the life of a child? Before you answer, what is the market value for the horrible disfigurement of your son or daughter? What is the market value of adding, say, 0.05 per cent toxic substance to your family's drinking water? How about 0.5 per cent? How about 1 per cent? These questions have no meaningful answers.

One of the great hallmarks of science is acknowledging the unknown and drawing only qualified and careful conclusions. The truth is that we simply do not know the future value of a rainforest. We do not know the full function of each species in an ecosystem, much less its monetary value. But, this doesn't stop practitioners of cost–benefit analysis from pushing forward and pretending to have absolute answers. Even when these unknowns are acknowledged, analysis will focus on conjectured numeric values. Meanwhile, important, but non-monetized, costs are simply ignored. So, what is unknown or difficult to quantify is presumed in practice to be zero. The most valuable and precious losses to our planet, our collective lives, or perhaps our future survival, may simply be presumed to be zero since they cannot easily be assigned a monetary value.

The noted geographer and analyst John Adams, himself a member of the Roskill Commission charged with a huge cost–benefit analysis to determine the location of a new London airport in the late 1960s, has likened this kind of analysis to a recipe for horse and rabbit stew.[18] The recipe details every aspect of preparation of the rabbit, taking great pains to describe every element of spicing, mixing, dicing, and so on. This is very much like the highly detailed measurement of tourist revenue, survey data, development costs, design specifications and so on. Then, the recipe continues, 'throw in one horse.' The resulting stew, of course, no matter how precisely and carefully prepared at the start, tastes a lot like a boiled horse. In short, this kind of analysis concentrates its focus on the things that can be measured and quantified. But, in the end, it is those things that cannot be so easily put in material terms that have the most importance.

Moreover, the application of so-called discounting means that short-term profits are almost always preferred over long-term ecological preservation. The logic of discounting is that costs are always greater if we pay them now than if we pay them later. If we spend the money now, after all, we lose the opportunity to earn interest on our funds in the future. This sort of analysis may make sense in comparing alternative financial investments, but what meaning does it have when deciding if you, your children, or your great grandchildren will suffer? Or whether a lake and its species will survive?

Here's how discounting works. Let us say we're evaluating a factory at the shore of a lake. The factory's emissions, if we do not install pollution mitigation devices, will make our lake completely dead and dangerous for swimmers by 2050. This lake might have a tourist flow that is estimated to generate, say, $2 million in revenue by 2050. Since we can't easily measure the ecological value of the lake or the value of the species that are endemic to the lake, they'll just be ignored. So, at a standard discount rate of 8 per cent, the lake's $2 million value in 2050 is valued at a paltry $37,000 in present dollars.[19] Since pollution abatement will cost $40,000, the only logical decision is to continue heavily polluting the lake; and now we can justify our choice with putative scientific certainty! This works the other way around as well: to justify an investment (or loss of profit) of even the meager sum of $2,000 now, we would need to be convinced of over $4 million of measurable monetary cost in environmental damage over the next century. In the end, addressing a problem early and averting a catastrophe is rarely cost-effective. By this logic, it's almost always better for our children and grandchildren to pay the cost than it is for us to do so.

It is this sort of tortuous logic that leads policymakers to argue that reducing greenhouse gas emissions is too costly while they disregard the overwhelming and certain human and ecological costs of global climate change.[20] US federal policy requires that all environmental policy decisions meet the criteria of cost–benefit analysis. Deciding whether or not keeping arsenic out of drinking water is worth the expense, or whether the benefits of clean air are worth the costs, comes down to the calculations of bureaucrats whose mandate is to decide the trade-off between public health and maximum economic return.[21]

It should be no surprise that the World Trade Organization insists on the use of such analysis. In fact, the WTO's rules push risk analysis logic to its moral limits. A country cannot restrict a product because the health effects are unknown or because it has not been proven to be safe; a product must be shown to be unsafe, in the narrowly construed terms of risk analysis, before it can be restricted.

For example, in 1996 a European ban on the import of meat from animals treated with growth hormones was challenged in the WTO by the US and Canada. The European Union had a long-standing series of bans on the use of such hormones within Europe as a result of widespread consumer concern over the harmful health effects of hormones. Rather than wait for a health crisis to emerge over time, Europeans chose to ban pre-emptively what was widely viewed as a dangerous practice. A WTO panel and later the Appellate Body found that the restrictions violated WTO rules because they werre not based on WTO accepted risk assessment techniques. Manufacturers, on the other hand, were not required to demonstrate conclusively their products' long-term safety.

Europeans argue that their obligation to protect citizens against the uncertain risks from hormones required the use of the precautionary principle. This alternative approach begins with the opposite perspective on risk analysis. The idea is if an activity or product raises a potential threat of harm to people or the environment, precautionary measures should be taken to ensure public safety, even before a final scientific conclusion is available. The aim is to avoid unnecessary risks with things as valuable as human life and the earth. Like science itself, the precautionary principle takes a wary approach to the unknown, and thus shifts the burden of proof. Until it is clear that a product is safe, restrictions will ensure the safety of the public and the environment. Advocates point to the numerous products – such as PCBs, DDT, and lead – that were thought to be safe, only to be discovered later to have serious effects following long-term exposure.

Some international agreements have recognized the need for a precautionary approach. The Cartagena Protocol on Biosafety, for example, states that the 'lack of scientific certainty ... regarding the extent of the adverse effects' shall not prevent appropriate action to protect human life.[22] But, as we saw in the previous chapter, it's not the formal rules that matter, it's the distribution of global power. The WTO, backed by the US and corporate interests, trump the Protocol on Biosafety. Risk analysis and its requirement of absolute certainty before protective action may be taken favor major industries and profits and therefore win the day. Indeed. US trade representatives ensured that a clause to the Protocol was specifically added to guarantee its subordination to trade agreements.[23] The issue is not reason, safety, or livability; it's power and profit.

So it is cost–benefit analysis and not the precautionary principle that defines national and international policies. Because cost–benefit and risk analysis use a complex set of specialized, self-referential terms and suppositions, it is extremely difficult for the average person to understand and participate in the decision-making process. For corporations and policymakers, this is yet another advantage, the capacity to defend their decisions in terms that intimidate the public and prohibit meaningful debate. So, while such analysis is neither objective nor scientific, it is efficient, if by efficient we mean able to quickly justify desired outcomes while excluding public discussion.

Modernization

Applied globally, this surrender to the dictates of the free market massively compromises the welfare of countries, communities, and ecosystems around the world. Neoliberal ideologues insist that the world's impoverished countries will ultimately benefit from this total capitulation to the market. Complete and unregulated exposure to global markets will foster industrial competitiveness, they argue. This in turn will lead to 'modernization.' By modernization, they mean the adoption of the social, political, and economic systems of the wealthy, as well as the acceptance of their cultural values and priorities. In short, this prescription calls for the complete embrace of the global system of production and competition as the solution – not the cause – of global poverty. If a country follows the path of industrialization and modernization blazed by the world's affluent countries, the neoliberal faithful assure us, they too could be wealthy.

Yet this prescription is troubling for a number of reasons. Even if we accept it on its own terms, the remedy for poverty requires enduring the horrors of child labor, mass paucity, disease, hunger, and dramatic environmental destruction in the hope that eventually your lot will improve. Exploitation and environmental destruction, in this perspective, are necessary steps to development. Moreover, most wealthy countries, the

4.1 / Increase in real per capita incomes, 1970–95 (%)

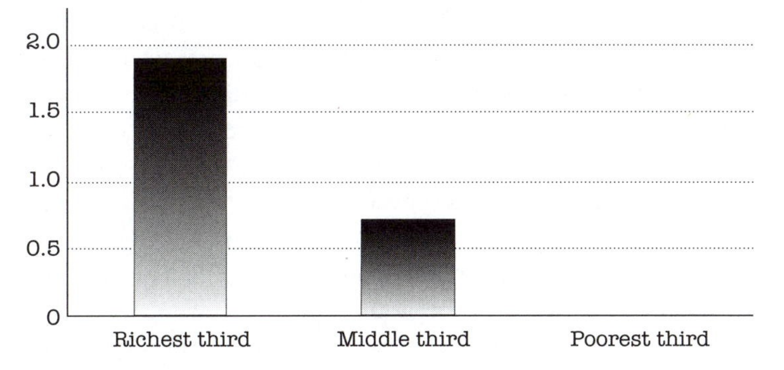

Source: United Nations Development Program, *Human Development Report 2003, Millennium Development Goals: A Compact among Nations to End Human Poverty*, Oxford University Press, New York, 2003.

4.2 / Questionable logic: the Environmental Kuznets Curve

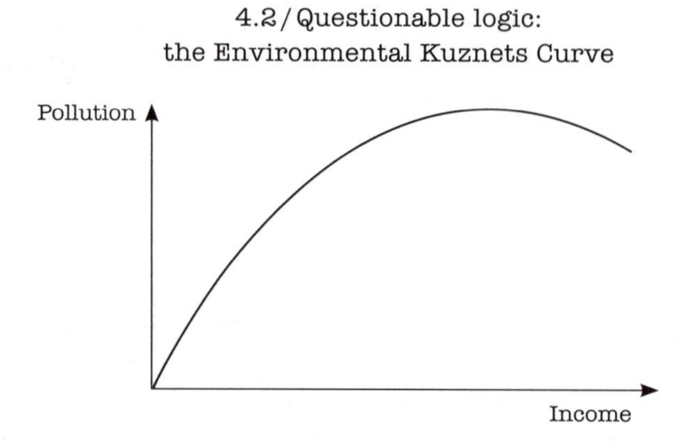

US included, built key industries by protecting their markets until they were strong enough to compete with foreign industries. IMF and World Bank policies focused on unrestricted markets specifically exclude this option. More importantly this prescription ignores the fact that the world's wealthy countries developed their wealth in large part on the back of the world's colonized poor. If not for the resources and labor extracted from colonies and dependencies, American and European economic history would have looked very different. In short, the world's poor are attempting to develop in a world where the winners and losers are already set, and the terms of trade, resource allocation, finance, and control are stacked against them. Incorporating themselves more deeply into the global system without fundamentally changing the way that system operates will lead only to further exploitation.

These free-trade evangelicals also claim that modernization will eventually reduce the rate of environmental destruction. They describe a so-called Environmental Kuznets Curve, arguing that while environmental impact initially increases greatly with industrial growth, as economies become more efficient, the environmental impact begins to decrease. However, recent studies indicate that this decrease is by no means clear.[24] And whatever damage was done on the way to this income level is not taken away once that level is achieved. Rivers that have suffered centuries of heavy metal dumping will not suddenly become clean when the level of dumping is reduced. Moreover, since incomes and wealth in many countries are in fact decreasing, a decrease in environmental impact seems unlikely in many places even if we buy the argument.[25]

Most crucially, this Kuznets Curve examines environmental destruction in a very narrow sense, most often measuring air quality or a similar local factor. But it's quite clear that in *absolute* terms, industrial growth

4.3 / Wealth increases environmental impact:
the relationship of per capita income and ecological footprint

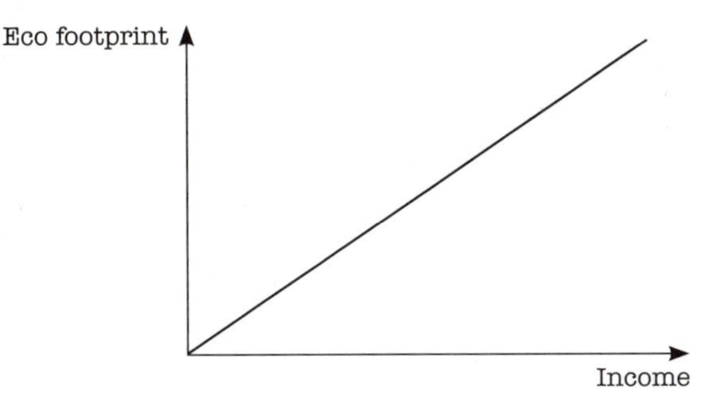

leads to ever-greater environmental costs. As consumption increases with increased income, the total demand on the earth's resources unavoidably grows. It may grow a little or it may grow a lot, but it grows. Thus when we look at the ecological footprint and income for countries around the world there is a clear pattern: the wealthier the country, the greater its impact on the global environment. Despite the supposed efficiency of US industries, the massive consumption of the US makes its citizens the most environmentally destructive in the world. As more countries expand their industries with a narrow fixation on producing more goods for more and more consumers, the total devastation will increase.

Hence global corporations and their allies in the world's most powerful international institutions argue for 'stability' – that is, don't rock the boat – and 'free trade' – that is, unrestricted access to the world's resources for the industries of the rich. And, they tell us, don't worry about the environment; it'll take care of itself if we just focus on making money. I'm not convinced. By giving corporations a free hand in their exploitation of the natural resources and cheap labor of the impoverished world, what these policies really mean is the continuation of a system that ensures the wealth of the wealthy few and the continued exploitation of the majority while wreaking devastation on nature.

This sort of logic allows American and European multinational corporations to undertake projects and activities that are so environmentally caustic and morally reprehensible that they would be simply beyond the pale at home. An infamous example is provided by the American mining company Freeport McMoRan's copper and gold mining in West Papua (called Irian Jaya by the occupying Indonesian state). Working at elevations of more than 13,500 foot above sea-level, the operation has decapitated the highest peak between the Himalayas and the Andes, a mountain that is sacred to the indigenous peoples of the island. Four hundred feet have been ripped off the top of the mountain. Within forty years, a full 3 billion tonnes of rock, trees, soil and other material – called overburden – will be cleared from the top and dumped over 2,000 acres of surrounding valleys and meadows. What will be left is a huge crater and a heavily contaminated landscape buried in 1,500 foot of waste. After a recent expansion of operations, output of contaminated rock and slurry waste – known as tailings – exceeds 100,000 tonnes every day, a number that dwarfs any other such mine. Much of this is being dumped into the local river, causing the river to breach its banks and spill across floodplains into nearby rivers and watersheds. At a minimum, the company expects an area of 50 square miles to be completely overwhelmed by the overflowing sediment and slurry. This sort of dumping has been condemned by the international community and is illegal in the United States and all European countries with significant mining sectors.

The local population has paid a dear price. An investigation by a US agency found 'major environmental, health, or safety hazards with respect to the river ... and the local inhabitants.'[26] Water contaminated with acid mine drainage and toxic metals has already caused skin rashes, stomach problems, bloody coughs, and death. The killing of local fish, wildlife, and plants threatens the livelihood of the local indigenous peoples. And, they have benefited very little from such so-called development; locals fill only about 2 per cent of the jobs connected to the mines.[27] Efforts at resistance by indigenous peoples have been brutally put down by an Indonesian police state that is bought and paid for by the US mining corporation.[28] Protection money paid to the military by the American company to squash local opposition has led to violent repression and untold deaths, including the recent deaths of two Americans.[29]

Neoclassical theory would try to convince us that this is a good thing, that Freeport McMoRan's rape of West Papua will lead to modernization and open the door to future prosperity. This is ludicrous. The people of West Papua are watching everything sacred to them being destroyed. They are desperately fighting a losing battle against corporate genocide. They don't need to be 'modernized,' they need to be respected.

Racing to the bottom

Local communities have few options. Trapped in a global system set up through imperial conquest, impoverished countries are left with little to sell but the labor of their people and the well-being of their environment.

While corporations select the next location of their factory, plantation, or mine, communities around the impoverished world vie for the dubious opportunity to be the next expansion site for the global system of exploitation. The result has been called a global 'race to the bottom.' As countries compete to attract foreign investors, environmental protections are ignored, rules protecting the rights of workers are disregarded, and wages are lowered, all to make the country more attractive to exploitative industries. The perverse reality is that to many around the world, it appears that the only thing worse than being exploited by global corporations is being ignored by them. With access to resources, credit, and markets completely controlled by the world's wealthy, desperately poor communities around the world are left with no choice but to contend for the favor of corporate interests. Thus the world's impoverished spend scarce assets attracting investments which will at best result in bare subsistence.

Grassroots struggles to defend the right to a decent life or to end environmental abuse can be seriously undercut by this process. If a strike for livable wages is launched in one factory, for example, production can simply be shifted to any number of other factories in other countries. The workers are fired, the plant is shut down, and the company's problem goes away. The result is ever-decreasing wages and working conditions.

A similar game is played with the environment, appropriately called 'slash and burn capitalism.'[30] Fruit producers in Central America, for example, use environmentally unsustainable farming techniques in one locale, degrading the soil, and polluting the ecosystem. After a few years, when the overuse of chemicals has destroyed the land's productivity and workers have begun demanding a livable wage, the company simply relocates, leaving a mess in its wake. Another company may then take advantage of the crisis conditions – unemployment and low land values – to begin yet another exploitative venture. The local residents are trapped in an unending cycle, with deep poverty and a heavily contaminated landscape as the result. And so it grows increasingly clear that the normal course of international commerce is fundamentally at odds with the preservation of the earth's ecosystem as well as our moral integrity.

Of course, the citizens of wealthy countries pay a price as well, as their environmental protections and the bargaining power of their workers are seriously undercut. In an effort to avoid environmental restrictions, heavily polluting industries have reduced their production in the US and Europe and increased manufacturing in impoverished countries where environmental regulations and public awareness are less stringent.[31] Chemical industries that utilize highly hazardous materials such as benzidine and asbestos, for example, have been particularly keen to migrate to countries with weaker environmental protections. Similarly, when US corporations chose to build two new power plants just a few miles south of the US border in the Mexican desert, it was absolutely clear that the freedom from US pollution controls was among the benefits.[32] Even when industries do not move operations to an impoverished country specifically to have a free hand to pollute, they are certainly willing to take advantage of the opportunity to lower their health and environmental standards once they relocate.[33] One result is pollution havens, highly polluted regions in impoverished countries that have attracted industries from wealthy countries.

Free-market economists debate the degree to which pollution havens really exist. They dismiss the notion of industries moving en masse to impoverished countries to escape environmental regulations as a fantasy of overly cynical environmentalists. While they readily admit that western-owned industries are often far dirtier than they would or could be at home, along with industry advocates they point out that Western industries are often more efficient and can be less polluting than local industries in impoverished countries. So, it's possible, they claim, that moving industries south could improve the local environment. But this argument understands the actual global impact of industrial migration much too narrowly.

The overall effect has little to do with whether this industry or that corporation moved to avoid environmental regulations or whether this factory or that one is more or less efficient. International competition has led to a subtle but ever-present climate of intimidation and desperation that works strongly against the development of stronger health and environmental protections in impoverished countries.[34] Public officials may be concerned over the environment, but they are far more worried about their immediate political futures, and these depend on their ability to court wealthy corporations. It's clear to all involved that enforcing rules too rigorously, or adopting new protection policies, makes it difficult to attract much-needed investment.[35] Policymakers are keenly aware of the pollution standards in competing countries and regions, and most are none too anxious to risk their career for the sake of the forest.[36] And, even if they were so inclined, industries would simply move to a country or region with weaker environmental protections, leaving the local population deeper in poverty and the global environment no better off. The results are pollution havens in some cases, and, overall, weaker environmental laws, more lax enforcement of existing laws, and a strong dampening effect on the creation of any new protections.[37]

The overall impact of corporate globalization, then, is clear: the environmental cost of ever-increasing consumption, principally centered in

Western countries, is being wrought on the peoples and landscapes of the impoverished world. In the end, the entire world is made a pollution haven, a toxic dump at the exhaust pipe of the global machine that feeds the rabid and increasing demand for more. Increasingly, because the world's rich are loath to pollute their own backyards, or because stronger environmental protections prevent them from doing so, this exhaust pipe discharges among the world's most vulnerable.

Commodified lives

The inevitable consequences of this sort of immoral market logic are visited on the world's most vulnerable peoples every day of their lives and in the most brutal of ways. Of course, treating people as commodities is nothing new. The very development of industry and an associated 'labor market' entailed putting a price tag on people's lives. Lives, like any other item on the global market, are bought and sold at whatever price the market will bear; and, I should add, global competition has seriously depressed the value of human beings.

Around the world, impoverished people toil in appalling conditions to produce goods for the Western markets. They make the items that can be found in any department store. A shirt may have been sewn by one of the countless women and girls in Indonesia who barely get by, working in the country's garment industry. Imagine the everyday life of any one of these workers. She may work 12 to 17 hours per day, often seven days a week, with no overtime pay. She has no choice; overtime is mandatory, and jobs are scarce. She may even be grateful: the extra hours may well be the only thing that allows her to buy food. New clothes, medicine, and other 'luxuries' are simply out of reach. She probably has trouble breathing at times; serious respiratory problems are one of many illnesses brought on by hazardous work conditions. The long hours in cramped quarters and awkward positions may leave her with chronic pain and fatigue, like so many other women at the factory. Like women workers in Bangladesh, Guatemala, and hundreds of factories around the world, she is regularly verbally and sexually harassed by her supervisor, on occasion beaten, and even threatened with rape.[38]

The examples are endless. Imagine the Mexican worker in the maquila industry just south of the US border. She is living in a virtual shack, surviving in extreme poverty, without indoor plumbing or electricity, barely able to afford food and other basic necessities. Again I say 'she' because roughly 80 per cent of textile sweatshop workers are young girls and women. Often, she is locked in the factory during work hours without access to a toilet. Armed guards may even be used to stop workers from leaving until they have finished their daily quotas. At times, this may mean working through the night, though workers may or may not be paid for the extra hours.[39] Accidents and injuries on the job are common, given the hazardous machinery and inadequate safety measures and equipment.[40] If she is injured and unable to work, her family would be unable to get by. If she misses work or is late, she may be made to pay a serious fine. Of course, any fine is serious when you live on about $30 a week. Like many garment workers, she may be from an indigenous group, which probably means enduring added abuse and mistreatment from employers and local officials.

Such conditions are facilitated by so-called free-trade zones or export promotion zones. Officially, these are designated regions outside of the normal customs border where foreign trade and other financial regulations are not applied. In reality, these zones are essentially designated regions where normal employment, social, and environmental standards and laws are not enforced and multinational industries have a free hand to do as they like. Mexico's maquiladoras were an early example, but there are many others. Some 30 million people work in nearly 850 free-trade zones around the world. As their numbers grow, so too do the legions of faceless workers caught in the global machinery.

Jordan provides a clear example of the effect. In 1994, the US inserted a provision into Jordan's 1994 peace agreement with Israel allowing the country to export products duty free to the United States, as long as 8 per cent of materials used in these industries came from Israel. Jordan's Qualified Industrial Zones became centers of export production, with 40,000 workers and sixty factories producing goods for US consumption. But neither the workers nor most of the owners are Jordanian. Most of the factories are owned and managed by business interests from China, Taiwan, Korea, India, Pakistan, or the Philippines. These factory owners import young and easily intimidated girls from South and East Asia, ensuring that their employees are totally dependent on the factory owner and thus allowing them total control over every aspect of their lives. Fewer than half the workers are Jordanian; about 90 per cent are women under the age of 22, and most of them earn about $3.50 a day.[41]

These young girls often live in company dormitories. Taken from their homes, separated from their families, and reliant on the company for their food, water, shelter, and every other aspect of their lives, these girls become totally dependent on factory owners. Workers live together in army-style barracks next to the factory, often six to twelve in a room.

4.4 / Child labor estimates

Africa	80 million
Asia and Oceania	153 million
Latin America	18 million
World	250 million

Source: ECSOC, *The Cost of Poverty and Vulnerability*, United Nations, 2001.

With a grueling work schedule, they rarely leave company grounds. If an employer decides not to pay for extra work, or to charge exorbitant rates for housing or food, there is little a worker can do. The conditions are atrocious; the profit margin is great. For a typical $23 sweatshirt, about 11 cents go to labor costs.[42]

When employers relocate to take advantage of still cheaper labor elsewhere, hundreds of workers can be left behind with no jobs and often owed weeks or months of back pay. Local authorities can and will do little. Since their visas restrict them to working only for the factories that brought them to the country, these people cannot work, cannot afford to eat, and cannot afford a ticket home. They are stranded in the no-man's land of the global economy.

These conditions are repeated around the world. In China, workers making clothes for Disney, Wal-Mart, Kmart, and other American corporations work 16-hour days, seven days a week in peak seasons. Workers can be made to pay serious deposits when they're hired, which are only returned after two years of work. This traps them in a system of virtual forced labor.[43] Children between 9 and 13 years old are forced to work 16-hour days with only small breaks. Shifts can routinely run to 2 a.m. and food rations are often reduced to save on costs.[44]

The excuse is often made that the variation in wages is not as great as it appears because of differences in exchange rates, living costs, or life expectations. Whatever the exchange rate or cost of living, human suffering is directly translatable across cultures and economies. Working 15 hours a day in hazardous, painful work translates perfectly, whether you're living in Michigan or Pakistan. Surviving in a shack with no toilet or safe drinking water, or living in overcrowded and unsanitary company barracks, does not need to be adjusted for the exchange rate. It is unacceptable and a violation of basic human dignity.

Labels marked 'Made in the USA' offer no certainty that your shirt or shoes were not produced under these brutal conditions. The Northern

Mariana Islands – a US Commonwealth in the West Pacific and thus part of the US – attracts factory owners who make clothing for some of the biggest brand-name labels. With complete access to the US market, no import tariffs or quota restrictions, low wages and lax immigration laws, the potential profit is significant. Young women from Asia are imported by factory owners. Lied to, intimidated, and desperate, these young women are actually made to pay a fee for their recruitment and relocation and made to sign a contract that traps them in indentured servitude by requiring them to work until they have paid off their debt. Since housing, food, and many other things can be charged against this account at any rate the factory owner deems appropriate, the debt is virtually endless. Denied their basic human rights, living and working in crowded, cramp, and unsanitary conditions, working up to twelve hours a day, seven days a week, these entrapped servants produce clothes for shopping malls throughout the US – all marked 'Made in the USA'.

However, you don't need to cross an ocean to find these conditions. Sweatshops in New York, Chicago, and around the US are all too prevalent. Taking advantage of new immigrants, some of whom may not have proper documents and thus be particularly vulnerable, unsavory entrepreneurs pay bare starvation wages for 15-hour days in grueling conditions. The tomato you eat may have been picked by an agricultural worker as young as 10 or 12, working 12 hours a day for far less than minimum wage. The work is punishing: weeding fields, picking cherries, packing watermelons, all while exposed to an oppressive sun and dangerous chemicals. The practice is detestable but perfectly legal. US laws have few protections for child farm labor. Hence the sons and daughters of impoverished Latino farm workers join in a struggle to feed the family. About half will finish high school; very few can hope for a better future.[45]

A growing number of oppressed workers around the world are fighting for their rights, but the battle is not easily won. Most won't join a trade union because they are intimidated and bullied by the supervisors or too scared of losing their only means of survival.[46] Working 60–80 hour weeks, they are all too exhausted to launch a seemingly hopeless struggle for better conditions. And the company's hired thugs can be brutal with those who challenge the system. If the factory owner has imported them, workers who try to start a union can be summarily deported. And a factory owner can simply shift production to yet another impoverished community if workers demand better treatment. So it's no surprise that labor unions are rare. In the hundreds of free-trade zones around the world, there are fewer than ten active unions; only two, in the entire history of these zones, have ever been in a position to negotiate wages and work conditions with management as the representative of all workers.[47] These hapless souls are fodder for global factories, virtual prisoners of the global blue jean, T-shirt, or running shoe industry.

Yet these are not the most extreme examples. Far too often, workers are not simply virtual prisoners of poverty; they are *literal* prisoners, slaves whose lives, labors, and futures are completely controlled by a slaveholder. Slavery is far more common now than it was at the height of the transatlantic slave trade. Even using a very narrow definition of slavery, only counting those whose lives and labor are for all intents and purposes totally controlled by a slaveholder, there are at least 27 million and perhaps as many as 200 million slaves in the world.[48] At a minimum, that's a population the size of New York, New Jersey, and Delaware combined.

In the Ivory Coast, for example, young boys are sold into bondage, made to work 12- to 15-hour days, beaten with sticks and chains, and otherwise physically and emotionally abused to feed the global demand for chocolate. These boys, aged between 12 and 15 and sometimes younger, are virtually kidnapped from their homes, forced to do the backbreaking harvesting and hauling of cocoa beans, and made to live in dire squalor with little food, all to produce a product, chocolate, which none of them will ever taste or even see. These plantations provide a full 43 per cent of the world's supply of chocolate. While not referred to as slaves by the plantation owners, since slavery is officially illegal, they are viciously beaten if they try to escape, and most will never see their homes or families again. When they are no longer able to work, they are simply tossed aside.

Bonded child labor is an expanding epidemic. Nearly 250 million children are trapped in labor around the world. The vast majority are farm workers. Those who are not in fact slaves have two options: work as told or starve. Children as young as ten are made to work from before the sun rises to ten at night. Often they suffer beatings, grueling conditions, a lack of food and water, and exposure to hazardous pesticides. A recent Human Rights Watch report found that in Egypt, Ecuador, and the US, children working in recently sprayed fields experienced headaches, rashes, nausea, and diarrhea.[49] In severe cases, pesticide exposure can lead to coma and death. In Egypt, children were made to work 11-hour days with only two breaks to drink contaminated water from the local canal. Like the forests, rivers, and air, these people are fodder for the profit-making machines of global production.

The extensive and brutal slave system that produces the charcoal that feeds Brazil's steel industry makes clear the link between the devastation of human beings and the devastation of the environment. Rainforests aren't the only thing being destroyed in Brazil; with the forests go the communities, traditions, and lives of countless impoverished people. Uprooted from their homes and traditional communities, workers find themselves caught up in a cycle of poverty and disadvantage from which there is little escape. Lied to, manipulated, threatened by agents of wealthy landowners, and desperate to feed their families, poor workers are tricked or forced into signing labor contracts or agreeing to contrived debts that they then must work to pay. They work, but, subject to unsavory bookkeeping, intimidation, and overpriced meals and shelter, the debt never decreases.

The conditions are atrocious. The workers are brought to camps isolated in jungle regions with no electricity, safe water, or communications with the outside world. They live in hastily built shelters that are less than shacks. Cut off from friends, family, and anything they know, penniless, and fifty miles or more from the nearest village, they are completely dependent on the local bosses. Employers can use violence at will to enforce their control or bleed more work from their hapless victims. They will spend the next several years at the camp making charcoal by hand. Waves of smoke, ash, and heat engulf the workers as they work the charcoal ovens. With no protective gear whatsoever, serious burns are common, and medical attention is unavailable. Scarred from countless untreated burns, suffering from serious respiratory illness, malnutrition, and exhaustion, after a few years, when all the surrounding forest has been gouged to a lifeless wasteland of ash reeking of smoke, they may be released. Of course, with no other options or hopes, they will probably end up in another such camp soon.[50] People who once lived in harmony with the surrounding forest are forced through desperation and violence to destroy it.

These are the workers who feed the smelters that produce the parts for our cars and homes. The US imports billions of dollars in car parts, construction supplies, and other metal products from these mills every year. Like the destruction of the forest, the labor of these slaves is entwined with a global order fixed on providing the cheapest possible products to wealthy consumers in the world's richest countries. In an extensive examination of slavery around the world, Kevin Bales made clear the link between the everyday lives of consumers in wealthy countries and the suffering of slaves: 'Across the world slaves work and sweat and build and suffer. Slaves in Pakistan may have made the shoes that you are wearing and the carpet you stand on. Slaves in the Caribbean may have put sugar in your kitchen and toys in the hands of your children.... They are paid nothing.'[51] They live on the margin of survivability. They are often chained to their worktable, forced to sleep next to their loom, locked in the factory. Their every waking hour is a work hour. The slaveholder controls all they are or own. They are cogs in the machine that feeds our insatiable demand for more and more stuff at ever-cheaper cost.

The result is an 'absolute glut' of potential slaves.[52] Slaves have become so cheap their use is expanding into many new sorts of work. They are so cheap that a slaveholder has no reason to protect them from injury or make sure they survive their enslavement – in fact, you could save money by letting them die. They are the 'disposable people' of the global economy.[53] The cheap and easy availability of slaves in turn brings down the already-miserable wages for non-slaves, since these workers must now compete with slave labor. Thus the cycle is complete: greater poverty fosters slavery, which in turn fosters greater poverty.

The beneficiaries of this system have no trouble justifying their actions; profits are high and that is all the justification that is needed. Like any other commodity, people are bought and sold at whatever price the free market will bear. Food, water, air, forests, and life itself can be bought and sold at whatever price the market dictates. Why should people be different from any other commodity? At the base of this spiraling race to the bottom are the throwaway lives of millions of desperate slaves.

Whether it is turning human beings into fodder for industry or transforming rivers into sewers, the 'efficiency' of the system is unquestioned. A now-infamous 1991 memo from Lawrence Summers, then chief economist for the World Bank, made the economic logic of the global system clear. The leaked memo was notable for the candor with which it reflected what was obvious but rarely explicitly stated: the only factors that matter are narrowly defined economic costs and profits. Neither the destruction of the environment nor human suffering enters into the equation. As Summers' deeply disturbing memo had it,

> I think the economic logic behind dumping a load of toxic waste in the lowest wage country is impeccable and we should face up to that … I've always thought that under-populated countries in Africa are vastly UNDER-polluted.... Only the lamentable facts that so much pollution is generated by non-tradable industries (transport, electrical generation) and that the unit transport costs of solid waste are so high prevent world welfare enhancing trade in air pollution and waste.[54]

Summers goes on to argue that the toxic qualities of this waste should not concern us since much of the population in the impoverished world will not live long enough to contract the deadly diseases caused by such pollution. In short, their low life expectancy and high infant mortality makes them a perfect place to dump our lethal waste! Summers faced neither criticism nor condemnation for his proposal. Quite the contrary: he would go on to serve as President Clinton's treasury secretary and then as the president of Harvard University. The existing global system is managed with a set of rules that makes such a morally bereft and ecologically suicidal proposal appear to make perfect sense.

The overall effect of this global system is despicable, but its economic logic, to borrow Summers' word, is 'impeccable.' Environmental devastation and brutal exploitation are imposed on impoverished people simply as a matter of established economic practice even if policies are not explicitly devised to do so.[55] Practices and priorities that put the most heavily polluting industries in communities that are least able to bear the risks, or that drive wages of the most impoverished down to starvation levels, are seen simply as 'good business sense' by the wealthy corporations, a 'necessary evil' by the leaders of impoverished nations, and basically 'the way things are' by poor communities who suffer the effects. Perhaps it is the way things are, but it is not the way things have to be.

Everything's for sale / 5

In a trite expression, people repeatedly insist, 'Money isn't everything.' We commonly take for granted the idea that a sense of community, the wonders of nature, justice, and a whole range of other values are more important than money. Still, as we continue to make these claims, we allow the commands of markets and our pursuit of ever more stuff to define the conditions in which we live and our very prospects for survival. We increasingly surrender what we say matters most to the dictates of a global market that is driven by the pursuit of what we say matters least.

In surrendering to the marketplace, we have degraded virtually everything to the status of a commodity. We have transformed the vibrant life of nature into a dead product to be bought and sold to the highest bidder. Even the value of human life is defined on a global labor market. And with so many lives available, they are offered at bargain basement prices. We have cast the earth and efforts to protect the earth as variables of commerce, subject to crude evaluations of economic costs and benefits. Forests, rivers, and lakes are not precious gifts of nature; they are products, allowed to survive only if their preservation is more profitable than their destruction. Most often it is not. We need to recognize the consequences of these choices.

Food as a commodity

At first glance, it seems like there must be a global food shortage. Some 15 million people die of hunger every year.[1] And, while there have been

5.1 / Some can't get enough, others eat too much
(% overweight and undernourished in the US and India)

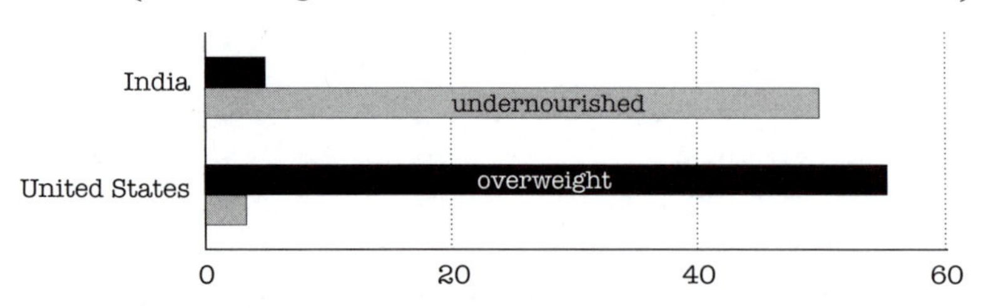

Source: Food and Agriculture Organization of the United Nations, *World Agriculture towards 2015/30*, Washington DC, 2003, www.fao.org.

improvements is some parts of the world, in Central and East Africa, India, Central America, and much of Asia, the ranks of the malnourished are increasing.[2] Some argue that the problem is population growth. The increasing number of people has outpaced our capacity to produce food, they claim. Either way, one could easily be led to assume that there is a gap between the number of people on the planet and the amount of food we can produce. However, this is simply not the case. As Frances Moore Lappé, founder of the Institute for Food and Development Policy, has it, abundance, not scarcity, best describes the world's food supply. If we only consider wheat, rice, and other grains, enough food is produced to feed every person on earth a 3,500 calorie diet each day. When we add meat, nuts, and beans, it's clear that the total amount of food produced in the world is more than ample. In fact, the total amount of food produced for each person on earth has increased. And it is not the case that more densely populated areas have more hunger. Yet, roughly 8 million people go hungry each day.

Why? The answer is simple. Food is not treated as a precious gift or a dear source of life to be cherished and shared. It is treated as a commodity, sold to the highest bidder. And the world's poor are increasingly finding themselves outbid, even for the food that they themselves grow. Of course, food had long been bought and sold. However, the global industrialization of food now means that this commodity is traded on a world market that is dominated by corporate producers and the demands of the world's wealthy. Food production is an industry; and, like any industry, it is meant to provide the wealthy with what they want, not the poor with what they need. So, as more and more of the world's farms produce for markets in the wealthy world, less and less is left for the impoverished majority. At the same time, the industrialization of food production is taking a severe toll on the environment. The toxic

5.2/Enough for everyone (global production of selected crops, calories produced per person per day)

Crop	Calories
Fruit	80
Nuts	53
Potatoes	572
Rice	638
Wheat	669
Other cereals	1,522
Total	3,534

Source: USDA (www.usda.gov) and FAO (www.fao.org) production data, calculated by author using average caloric values for each crop.

waste of this industry chokes our wetlands, rivers, and oceans; and the insatiable demand for more production contributes to the destruction of the world's soils and forests.

It's not overall scarcity of food but unequal access to food that ensures that widespread hunger persists. Although 43 per cent of the population of sub-Sahara Africa is malnourished, the continent produced enough food to feed the entire population an adequate diet. More than three-quarters of all malnourished children live in countries with a food *surplus*.[3] Economists would presumably call this a 'market failure' – a rather sanitized term for a practice that causes the hunger of billions and the deaths of millions every year while food rots in warehouses and silos around the world.

It's most often the policies and economies of human beings, not natural disasters or environmental limits, that cause even the worst examples of hunger. In Bangladesh in the mid-1970s, for example, severe flooding resulted in widespread hunger – not because of a shortage in food per se, but because flood damage resulted in mass unemployment and the doubling of rice prices, so people could not afford food. As a result, 1.5 million died of starvation when no actual food shortage existed. In fact, production of rice in Bangladesh was higher than it had ever been![4] Similarly, severe famine in West Africa in the early 1970s was commonly attributed to drought. However, plenty of water seemed available to grow the massive amounts of cotton, peanuts, and other exports during the period. In the midst of the reported drought, exports of agricultural products, largely to well-fed customers in wealthy countries, actually increased.[5]

5.3 / Global hunger: child malnutrition and overall undernourishment, 2000 (%)

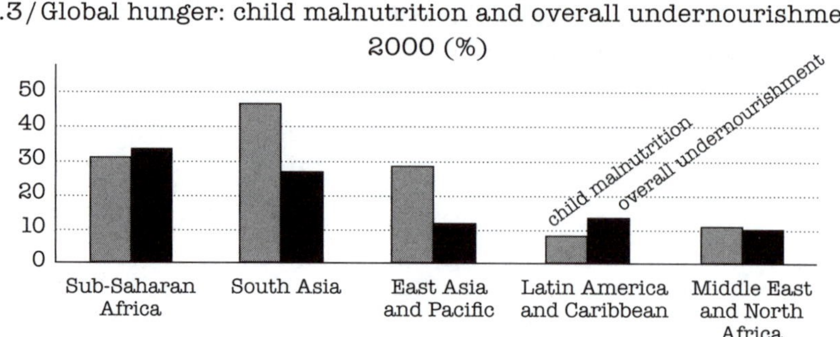

Source: *Millennium Development Goals*, World Bank, 2003; and *World and Regional Trends 1990 and 2000*. United Nations Statistics Division, 2003.

The problem is clear and becoming more pressing: global industrial agricultural production prioritizes crops that promise the most profit. Consequently, low-cost crops in poor countries are replaced by cash crops for sale in wealthy countries. Farmers who once produced staple crops for local consumption are displaced by large-scale agricultural industries that grow flowers, sugar, shrimp, beef, coffee, soybeans, and other luxury foods for the world's wealthy. Corporate farms take over the countryside and fewer and fewer small farmers have access to land. The perverse effect is that impoverished countries that export more food can actually be more likely to suffer from malnourishment.[6] Similarly, countries that produce more food can have higher rates of malnutrition. On the other hand, social inequality in a country is a clear predictor of not only low food availability but also low life expectancy and malnutrition. The conclusion is evident: we have plenty of food, but the poor can't afford to eat.

This is not only morally intolerable but also environmentally unsustainable. The massive use of chemicals and machinery in corporate farming promotes soil erosion, water contamination, the death of wildlife, and a public health threat. In this century alone, for instance, the US has lost half its topsoil (the fertile upper layer of earth necessary to grow all food).[7] Around the world, about 70 per cent of the land used for agriculture is at risk of being turned to degraded waste soil that will be worthless for farming, unable to sustain healthy and diverse flora, and unlikely to recover.

Industrial agriculture is also heavily water intensive. It is clear that when multinational corporations expand their control of land in a country, the use of water for agriculture increases significantly.[8] Because industrial

5.4 / Children pay the price for food inequity (1990)

	% of children		
	wasted	stunted	underweight
REGION			
Developing countries	9	41	34
East and Southeast Asia	5	33	24
South Asia	17	60	58
ECONOMIC GROUP			
Middle- to high-income countries	6	29	22
Low-income countries	10	45	38

Source: Food and Agriculture Organization of the United Nations, *The Sixth World Food Survey*, FAO, Rome, 1996, www.fao.org.

production for the global food market often promotes crops inappropriate to local climate and soil, such agriculture can have devastating impacts on the local ecological balance. Aquifers are overdrawn, rivers and lakes are depleted, and soil can be seriously degraded. All water contains some salt. The massive irrigation required by industrial agriculture, especially in inappropriate climates, can leave damaging salt residue. The soil progressively degrades and eventually is made unusable for farming. About 20 per cent of the world's farmland is affected by rising salinity; and about 2.5 million acres are abandoned each year as a result.[9]

The link between the expansion of industrial agriculture and the growing use of pesticides is also clear.[10] These pesticides poison the global ecosystem in ways we cannot fully identify and they endanger the lives of millions. Pesticides do not discriminate well; they kill beneficial predators and parasites as well as crop-damaging herbivores.[11] Thus the web of life near croplands can be virtually eradicated, leaving a devastated landscape.[12] Insects that are vital to the pollination of crops and wild flora are killed.[13] Rivers, lakes, and water tables are made toxic.[14]

The adverse effects of pesticides on birds, reptiles, and other wildlife are clear and overwhelming.[15] Many of the pesticides now in use are more toxic to birds than DDT. They also affect wildlife indirectly, by killing prey and vegetation needed for their survival.[16] Fish populations far away from the initial site of pesticide use can be seriously impacted.

Wildlife isn't the only thing being poisoned. In the US alone, the EPA estimates that 300,000 farm workers suffer from pesticide poisoning each year. And because pesticide poisonings are unlikely to be reported or treated, the actual numbers could be much higher.[17] Overwhelming evidence

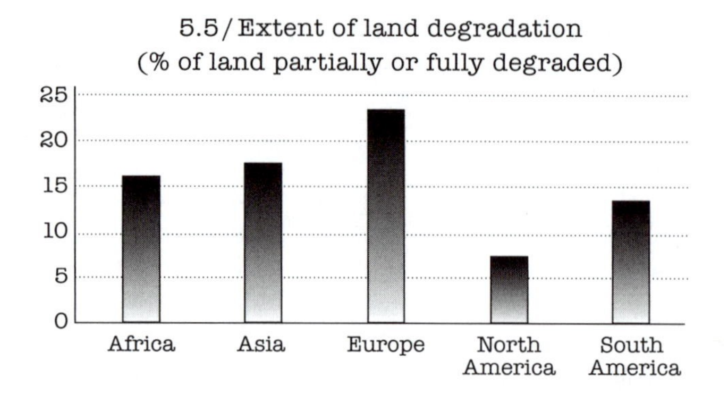

Source: United Nations Environment Program, *Global Environmental Outlook 3: Environment for Development: Past, Present and Future Perspectives*, 2002, www.unep.org.

indicates that increased risk of leukemia, testicular, liver, pancreatic, lung, and brain cancer, immune dysfunction, infertility, memory loss, mood disorders, and other neurotoxic effects may be linked to even low exposure to these chemicals.[18] Farm workers exposed to pesticides experience genetic damage, stillbirths, and abortions at much higher rates.[19] A recent study found that farmers exposed to organophosphates were almost six times more likely to suffer depression than those who didn't use these pesticides.[20] That much we know. What we don't know is what other health and environmental risks might be linked to the massive use of these deadly poisons, even for those who have never set foot on a farm.

It's clear that it is not just farmers who are exposed to the risk. More than fifty pesticides that are classified as carcinogenic are applied to food in massive quantities. More than a third of food purchased by US consumers has a detectible level of pesticides.[21] The US Department of Agriculture reports that perhaps half the groundwater and well water in the US is, or could be, contaminated by pesticides.[22] When the US Geological Survey conducted its National Water Quality Assessment they found that more than 95 per cent of the samples collected from streams were contaminated with pesticide, although the levels often did not exceed established drinking water criteria.[23] Yet for many chemicals, there are no set criteria; and when there are, the effect of cumulative exposure to the pesticides is not considered.[24] Thinking of giving up water? Pesticide contamination can reach into virtually every category of food production. Cows' milk and even human breast milk can be contaminated by pesticide residue.[25]

Despite these concerns, millions of tonnes of pesticides continue to be dumped on the environment each year. About 500,000 tonnes are used in

5.6 / Top ten agricultural uses of pesticides in the US (million pounds of pesticides/year)

Crop	Pesticide use
Corn	240
Soybeans	74
Cotton	72
Potatoes	60
Grapes	56
Citrus fruits	34
Tomatoes	27
Pastures	25
Tobacco	24
Peanuts	24

Source: *Agriculture, Pesticides and the Environment*, Organization for Economic Cooperation and Development, Paris, 1997, p. 55.

US agriculture and industry, and the EPA reports that an additional 3.5 million tonnes are used in wood preservation, disinfectants, and sulfur compounds every year.[26] Unsustainable crop management practices, rising resistance in pests, and the indiscriminate nature of pesticides that kill helpful as well as harmful species means that more and more pesticides are necessary to achieve ever dwindling effect. So, while pesticide use rises rapidly, crop loss due to insects is also increasing. In the four decades following World War II, corn losses due to insects increased fourfold, despite a thousandfold increase in insecticide use on corn crops.[27] Overall, the use of insecticide in the United States increased tenfold while crop losses from insect damage doubled.

Yet the US has it relatively good! The most harmful chemicals are used in the impoverished world. The world's wealthy countries have dumped over 500,000 tonnes of restricted, unusable, or unsafe pesticides on impoverished countries around the world.[28] The most dangerous of these, chemicals banned in the US, are being produced by Western industries in impoverished countries.[29] Don't think that you're safe just because you don't live in Africa. Pesticides used in Africa appear in Key West, Florida, a few days after spraying. Pesticides applied in Lubbock, Texas, turn up 1,500 miles away in Cincinnati, Ohio.[30] Distance is no protection from the expanding poison clouds.

5.7 / Growning world fertilizer use (kilograms per person per year)

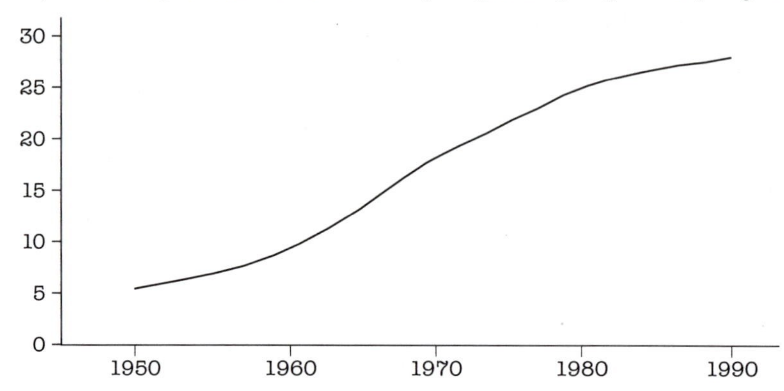

Source: Lester Brown, 'We Can Build a Sustainable Economy,' *The Futurist*, July–August 1996.

Nonetheless, it's clear that the world's poorest farmers face the greatest risk. Because protective clothing is not available and safety precautions are often not used or known of by uneducated farm workers, the health risks are great. About 25 million pesticide poisonings occur around the world, and at least 20,000 die from accidental exposure, every year.[31] If we include intentional self-poisonings, which may be linked to chronic physiological and psychological effects of pesticide exposure, that number increases more than tenfold.[32] Since much of this pesticide is used not to produce food for the local population but in the production of export crops for wealthy countries this is particularly disturbing.

While these techniques may produce greater profits, they are not better at producing food. The National Research Council has found that small-scale, more traditional farming – so-called alternative farming – requires fewer chemicals, and therefore lower costs, and also fewer antibiotics in livestock rearing than industrial measures, and it can be more efficient.[33] Small farms almost always produce greater output per acre. If we include environmental and social costs in our calculations, industrial agriculture would be exposed as even less efficient. Because wealthy industries can afford more chemical inputs, higher transportation costs, and massive land use, in the short term they enjoy greater outputs and higher profits. Of course, the growing tab on the earth is as yet unpaid, and at some point we're going to run out of soil to exhaust. Hence, although in terms of total energy used, environmental costs, and food produced per acre, these industries are less efficient than more traditional farming, since most of the costs are not paid – and profits are high – chemical-intensive industrial agriculture grows ever more dominant.

5.8 / Farmers' declining share of food dollars (%)

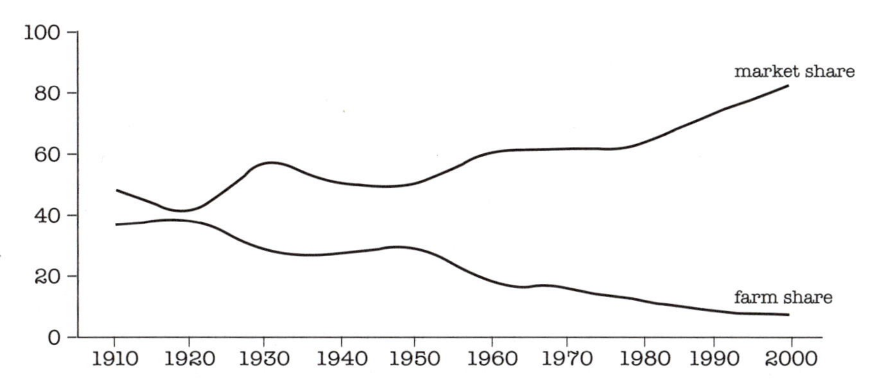

Source: Brian Halweil, 'Where Have All the Farmers Gone?' *World Watch*, September/October 2000, pp. 12–28.

The chemicals, mechanized harvesters, tractors, irrigation equipment, and seeds brought to impoverished rural areas by agricultural industries have dramatically increased the costs of farming, driving many local farmers from the market. Those who are able spend an increasing amount of their income on these expenses. So farmers are caught in an ever-tighter squeeze; they pay more to international corporations for the chemicals and machinery needed to keep farming, while increased competition from global agriculture industries lowers the price they receive for their crop. As a result, while the US population more than doubled in the twentieth century, the number of American farmers has decreased from over 7 million to about 2 million. The impact on small farmers in the impoverished world has been even more dramatic.

The overall result has been expanding corporate control of the world's farmland. In Central America, multinational corporations own up to half the agricultural businesses. By the end of the 1980s, in many countries, virtually all the plantation exports of luxury foods were controlled by multinational industries. Meanwhile, the production of rice, beans, corn, and other staple foods for local markets has dropped significantly.

When small farmers are not completely driven from the land, they are nevertheless taken over through 'cooperative' contracts with international corporations. Farmers are bound by an exclusive contract with a food producer who provides seeds, credit, and detailed requirements for the production of a designated export crop. This sort of indirect control of farms can be more profitable for industries since they receive a large share of the profit but are required to take few of the risks. The benefits to farmers are far less clear.[34] If the crop fails, the farmer endures the loss; if the crop succeeds, industry enjoys a big chunk of the profits. While nominally independent, these farmers in fact become disguised wage laborers whose independence is completely illusory.[35] Incomes can plummet to a few dollars a day. As their costs increase and their income decreases, many farmers find themselves unable to afford food for their families. Children are often pulled out of school and put to work in the fields in a desperate attempt to make ends meet.

Thus the corporate takeover of global agriculture works at both ends: redirecting food flows from the hungry to the rich, on the one hand, and increasing hunger by increasing landlessness and poverty on the other. The result is always tragic and often violent: mass hunger in rural areas that once produced surplus food; local communities, traditional cultures, and sustainable farming practices destroyed. Local henchmen, police, and military are brought in to clear farmers and peasants off the land. Wealthy landowners use intimidation and violence to enforce their control of the countryside.

India is one of many places where farmers are trapped in the resulting cycle of poverty, debt, and dependence. Farmers who had once grown a diverse set of crops for their families and a bit extra for the local market have had no choice but to place their lives and livelihoods in the hands of a global, corporate-controlled market. A few years ago, Indian farmers switched to chemical-intensive and water-intensive growing of genetically engineered cotton, convinced by international corporations like Monsanto that this would bring them prosperity. But, when the global cotton market dropped and pests developed tolerance to the powerful chemicals used, many farmers found themselves on the losing end of the gamble. While Monsanto enjoyed mounting profits, heavily indebted farmers watched their crops and livelihoods dwindle in value, and indeed size, by the attacks of pesticide-resistant worms and caterpillars.[36] Seeing no hope left, hundreds across the country took their own lives, many by drinking the very same pesticides that once seemed to promise prosperity.

Similar stories can be told of countless communities around the world: social instability, hardship, and mounting malnutrition and hunger; the loss of a precious and irreplaceable inheritance. With the destruction of small farmers comes a destruction of sacred legacies, cultures, and traditional social relations. Food is not a product like televisions or automobiles. Growing food has long been central to our shared welfare – a pillar of life, family, and community – and deeply bound up with sacred meaning. Food is our most direct connection with the earth, the seasons, and nature's prosperity. We sever that connection at great cost.

You could ask the obvious question at this point: why? Why are we using inefficient, dangerous, and destructive methods to produce our food when alternatives are available? Why do billions suffer with undernourishment in a world of plenty? The answer is not complicated: profit. Industrial production methods return the greatest profits for multinational food producers. Food is a commodity. Its production is not determined by a desire to feed the hungry, make the most healthful and nourishing food possible, or preserve the precious ecological balance of the countryside for future generations of farmers. The aim of global food industries is to make money; and this perverse, socially and environmentally devastating mode of production allows the greatest profit. For the global food industry, no other justification need be sought.

Plants and animals as commodities

A lot is never enough. Having redefined the production of food to ensure massive profits at the expense of human lives and the environment, multinational industries now hope to own the very definition of food. Think of the potential revenue: if you own the definition of a food product, say rice, you get a cut of the profit even when you don't grow, harvest, transport, or sell the product yourself. So being able to patent the biological definition of a crop would mean that farmers around the world would have to pay royalties to a multinational corporation to grow a crop even if it is sold only in local markets. The potential corporate earnings are almost limitless.

Genetic modification (GM) can make this possible. In essence, genetic modification is the process of creating a new form of life by putting one gene or more from one plant or animal into another. The process transforms a food crop, animal, or micro-organism into a variant that is not normally found in nature. For example, a trout gene can be put in a tomato, a mouse gene can be put in a potato, or a human gene can be put in a chicken. The new life form can then be patented; and the corporate producer can demand payment from those who use it. Advocates say that this will improve nature. Tomatoes can be made to stay red longer. Corn crops can be made to produce a poison that kills insects. Chickens can be made with extra large breasts. For now, GM is largely limited to crops for food, feed, and pharmaceuticals. However, work on genetically engineered trees, fish, and even bugs is under way.

When industries insert a gene or modify a seed, the new seed is legally a discrete and patentable commodity. A genetic 'marker' can be inserted with the new gene to identify clearly the proprietary property. Later generations of that seed may not be saved and planted without paying the 'owner' of the seed. The kernel of life, the source of food, is thus recast as a complete commodity to be bought and sold in the marketplace by a single owner.

Some opposition to genetic modification rests on a moral argument, rejecting scientific efforts to 'play God' or presume the capacity to improve upon nature. The more adventurous excursions in the biotechnology industry disturb many and raise fears of science out of control. Current research is going well beyond tweaking a species by adding a gene here or there. Research is rapidly expanding into the creation of entirely new strands of DNA.[37] While noting the value of science, Bill McKibben argues that we quite simply need to be willing to say 'Enough.' 'Sometime in the next few years, a scientist will reprogram a human egg or sperm cell, spawning a genetic change that will be passed down into eternity. We are sleepwalking toward the future, and it's time to open our eyes.'[38]

Critics argue that genetically modified food may present a serious health threat. For example, there exists the potential to create new allergens or unexpected toxins. While no adverse health effects have been proven, it is too soon to determine if any long-term health risks exist, so most reasonable researchers advise caution. The orthodoxies of risk analysis preclude such caution, of course. It makes sense to move cautiously until we are certain that GM foods present no public health or environmental threats. Instead, corporations are given a virtual free hand until the long-term hazards become absolutely clear, but by then it may be too late.

Most scientists are proceeding with caution and care. This is more than can be said for the corporations that end up controlling their creations. With huge profits to be made, corporations have been anything but cautious, manipulating science and deceiving the public for the sake of preserving their potential earnings.[39] In less than a decade, the acreage dedicated to genetically engineered crops in the US alone soared from zero to over 70 million. Just seven years after its introduction, over three-quarters of all soybean and one-third of all corn grown in the US is genetically engineered.[40] Well over half the products in any given grocery store contain some genetically altered ingredients.[41] Since industry has ensured that these products are not labeled, we have no way of knowing which products are genetically altered. Critics point out that this tremendous expansion over such a short period entails tremendous risk. Research, overwhelmingly funded by a few corporations in private labs rather than public research labs and universities, is moving furiously in the search for blockbuster

5.9 / The increase in GM crops (million acres)

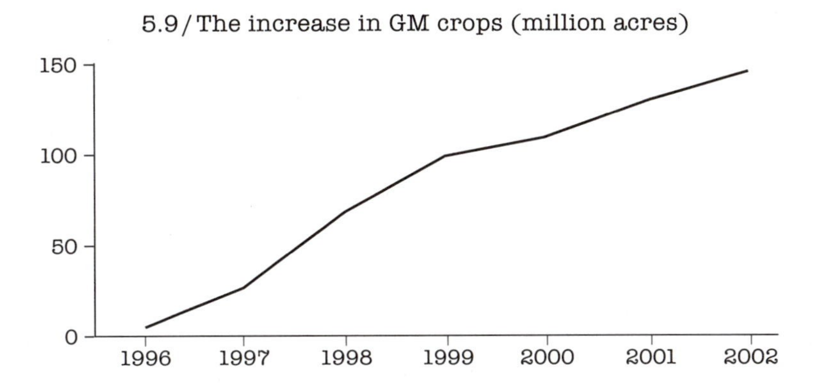

Source: *Global Review of Transgenic Crops*, International Service for the Acquisition of Agri-Biotech Applications (ISAAA), Manila, Philippines, 2002.

products that promise big profits.[42] We're playing with fire; and by the time we realize the danger it may be too late.

Biotechnology companies, and their government supporters, are betting a lot on the hope that no unforeseen public health consequences will emerge in the future; although neither they nor their critics can be sure of the odds. What is certain is that the biotech industry has been quick to squash those who dare to raise such concerns. When researcher Arpad Pusztai dared to suggest that GM potatoes might have harmful effects on laboratory rats, the reaction was severe. He was rapidly removed from his position, legally gagged, and soundly discredited. When fellow scientists protested, Dr Pusztai was allowed to speak again, but the damage was done.

Much subsequent research has cast doubt on some of Pusztai's findings, though some studies have also supported his conclusions.[43] But the validity of his research isn't the only issue. Further scientific research will tell whether he's right or wrong. The most troubling part of the story is the speed and harshness with which he was vilified and silenced for discussing potential health problems with GM crops. There's evidence that government and corporate agents colluded in Pusztai's case and others to muzzle unsupportive scientists and restrain unfavorable research results.[44] It's apparent that there is precious little room for an open debate on the health risk of GM foods.

The expansion of GM crops will likely accelerate the already critical loss in crop biodiversity brought on by the growth of industrial agriculture.[45] Over the past century, 86 per cent of apple varieties, over 90 per cent of corn varieties, 93 per cent of lettuce varictics and 80 per cent of tomato varieties have been lost.[46] In India, most rice production utilizes only ten of the many thousands of rice varieties available. Over a thousand varieties of potatoes exist; only four are widely grown. In China, only 10 per cent of the wheat varieties grown half a century ago are still cultivated.[47] The facilitation of monopoly control through genetic modification will intensify this trend, increase genetic erosion, and add to the threat to the sustainable production of food.

The dangers entailed in this loss are great. Plant breeding depends on a natural variety of germ plasm, the 'raw materials' of genetic variation. The relatively narrow strains created through breeding are especially susceptible to disease, so breeders must be able to return to the parent strains to ensure genetic variability. However, as newly created strains begin to dominate global markets, we risk reducing the original varieties and thus diminishing the essential source of new crops.

At the same time GM will only add to the burden of small farmers and agrarian communities. At the moment, most seeds around the world are freely shared among farmers. Local farmers exchange small quantities of improved crops, breeding them with local varieties to boost crop yields for subsistence and local sale. In India and Africa as much as 80 per cent of seeds are saved from previous crops. Overall, some 1.4 billion people depend on such saved seeds.[48] In many cultures these seeds, and the communal exchange itself, have sacred importance and a central material, cultural, and spiritual role. Through a clever maneuver of genetic engineering, biotech industries are planning to end such traditional seed sharing. Monsanto and Zeneca have developed a so-called terminator gene that can ensure that genetically engineered plant seeds are sterile; hence that the world's farmers will have to keep coming back for more and paying the price. Since it is large landowners, corporations, and global markets that largely determine what crops will be planted, small farmers will have little choice in the matter. This will ensure that even the smallest farmers are dependent on the multinational biotechnology industries that control the seed market. And by promoting the end of traditional crop varieties, it will place growers in an even more vulnerable position if the engineered crops fail or are overcome by disease. As well as farmers and activists, genetic researchers are concerned. The world's largest agricultural research group, the Consultative Group on International Agricultural Research (CGIAR), has called on research institutes to ban the use of the terminator technology.[49]

Proponents insist that GM crops can ensure less environmental damage. One justification often cited for these crops is that, since they can be made to produce their own pesticide, they will require lower applications of environmentally harmful chemicals. However, the opposite may in fact be the case. With crops engineered to withstand massive doses of toxic herbicides, growers use *more* of these environmentally destructive chemicals.[50] Herbicides used in massive quantities on herbicide-resistant crops can create a virtually dead field, free not only of undesirable weeds but also of the insects on which wildlife and the ecological balance depend. Of course, the increased run-off will kill wildlife and destroy water sources far beyond the area farmed. In theory, there is the potential for lower pesticide usage with GM crops. Yet the benefits of genetically produced pest protection could well disappear within a few years as insects develop resistance to the genetic alteration. This could lead to crop losses or, once again, high pesticide use.[51]

This indirect effect may not be the only environmental consequence of GM crops. The impact on wild flora from potential crossbreeding with GM crops is of concern. Plants engineered to withstand herbicides can easily produce fertile crosses with common weeds.[52] Unlike chemical contamination, genetic contamination can multiply itself, making even a small leak very difficult to contain. Despite industry claims to the contrary, recent research indicates that genetically engineered crops can out-compete wild species.[53] The result could be the eventual extinction of wild plants in favor of genetically modified competitors.[54] The impact on the evolution of pests and the effect on non-target species might be grave.[55] What's more, since most genetically modified crops have been engineered for human consumption, their effect on the wildlife that also eats them is not known. For example, we know that the pollen from one variety of genetically modified corn is deadly to monarch butterflies.[56] While the vast majority of corn now grown in the US does not produce this pollen, the fact that this effect was not discovered until after this corn was in cultivation underscores the risk. What saved the monarch butterfly was the fact that this particular variety of GM corn did not sell well. As a report by the Union of Concerned Scientists has it, 'it was just a lucky break – not government vigilance – that protected the monarch butterfly.'[57] Clearly the potential for other unrecognized risk to wildlife is high.

The threat of contamination is most troubling with the expansion of crops that are genetically altered to produce powerful drugs. Recently, US Department of Agriculture inspectors found in a batch of soybeans destined for human consumption a gene engineered to produce a drug for the treatment of pigs with diarrhea.[58] Just two month earlier, the same biotechnology company, ProdiGene, had been ordered to burn 155 acres of corn when a stray gene escaped containment. Yet this is only one case; more than 300 open-air fields covering thousands of acres are being used to grow crops that function as factories for powerful and potentially dangerous chemicals. These crops may well cross with crops grown for food and result in dangerous chemicals entering the food supply. Proponents point out that this research may lead to inexpensive vaccines and other medicines that could be made available throughout the impoverished world. However, critics rightly respond that all of these drugs can be produced through safer means. It is economics that keeps them from the world's needy.

Containing the spread of GM crops is already a problem. Genes from genetically altered corn have even been found in the local varieties of corn in the isolated village of Capulalpan, Mexico, for example.[59] This is particularly troubling since this Mexican region is the heart of the world's repository of corn genetic diversity. When disaster strikes corn anywhere in the world, farmers turn to this region and the nearly one hundred different varieties of blue, yellow, black, and white corn that grow there to diversify their crops and adapt to challenges from pests or climate. Hence the contamination of this irreplaceable source of crop diversity is deeply worrying; and it's not at all clear if these altered genes can ever be bred out of the population.

In 2000, Kraft food was forced to recall more than 2 million boxes of taco shells when it was discovered that they were made with a genetically modified corn that had not been approved for human consumption. As the crisis snowballed, it became evident that Aventis CropScience, the creator and distributor of the Starlink corn, in the words of an EPA senior official, 'blew it.'[60] The EPA had insisted that this corn be strictly limited to animal feed and industrial uses. Yet there were no records indicating which farmers, or even how many farmers, were producing this crop; no measures in place to ensure control of the crop; and no way of tracking the sale of the crop. Nevertheless, under pressure from powerful biotechnology interests, Congress has opted for the continuation of 'voluntary' regulations.

Biotechnology companies have been less than keen on containing this spread. Perhaps because the spread of GM crops, intentionally or otherwise, may mean that in the end all farmers will have to pay biotechnology companies royalties. An important precedent was set when the Canadian Supreme Court recently found that a farmer whose fields were contaminated with a genetically modified corn produced by Monsanto would have to pay the corporation for the seeds. When GM plants with proprietary Monsanto genes were found on his land, Percy Schmeiser, a Saskatchewan farmer, claimed that he never purchased or planted these crops. Perhaps their presence was the result of cross-pollination from genetically altered crops near his land, he suggested. The most striking fact is that the High Court agreed with Monsanto's claim that whether he planted the crop or not was irrelevant. Even if the appearance of the GM crop was the result of accidental pollination from wind, even if he didn't know the seeds were there, Mr Schmeiser must pay Monsanto for possession of the proprietary crop. This set a dangerous precedent. Far from being punished for the contamination of surrounding farmland, biotechnology corporations must be paid for allowing the uncontrolled spread of their crops!

It should be no great shock that biotech companies have encountered a serious public relations challenge. Their desire to overcome this image problem was clear in their recent push for so-called 'golden rice'. This rice is genetically modified to produce betacarotene, a nutrient that the body can use to produce vitamin A. Researchers hoped that this could help eradicate xeropthalmia and blindness caused by vitamin A deficiency. Suffering from a series of public-relations failures and growing public skepticism, biotech companies latched on to this development. In a series of print and television advertisements aimed at American heartstrings, and replete with soft-focus images of children and mothers, biotech companies promised to save 20 million children a year by distributing this new miracle rice. Even the Rockefeller Foundation, which financed the original research on golden rice, declared that the 'public-relations uses of golden rice have gone too far.'[61]

What the biotech firms did not say is that a child would need to eat about twenty-seven bowls of this rice a day to meet minimum requirements of Vitamin A.[62] Furthermore, betacarotene will not be adequately absorbed by the human body unless essential fats and proteins are available – but these are precisely what a malnourished child is missing. Of course, many of those suffering from vitamin A deficiency are already surrounded by abundant sources of vitamins and minerals and the capacity to grow much more; they just can't afford the food.

Certainly, the well-intentioned scientists who worked on golden rice are not to blame. They hoped that this and future research might lead to improved health conditions around the world. And perhaps it will. But the corporate panacea painted by Monsanto has little to do with this kind of noble aim. If biotech firms really wanted to help children, they could simply provide vitamin A supplements to malnourished families, with far greater effect. For that matter, making simple, wholesome brown rice available to impoverished communities around the world would go much farther toward eradicating malnutrition than the insertion of a gene to make a super rice that will help no one – no one, that is, except a dubious industry struggling to improve its public image by claiming the moral high ground.

Genetic engineering isn't the only way to reap profits from the ownership of life. The same intellectual property laws that allow for ownership of GM crops can enforce corporate ownership of non-GM crops. A classic case began in 1997, when the Texas-based corporation RiceTec was granted a US patent for basmati rice. Grown across some 4 million acres in South Asia, this rice has developed an international reputation and demands premium prices in both domestic and international markets. In India alone, the crop contributes hundreds of millions of dollars to export earnings. By usurping control of rice varieties bred by generations of farmers in India and Pakistan, this Texas company hoped to capitalize on this extremely profitable market. It hoped to own the right to use the term 'basmati'; and in the process it would effectively kill one of the few opportunities for profitable farm exports in the impoverished world. Similar predatory patents are threatening the control of Mexican yellow beans, jasmine rice of Thailand, quinoa and turmeric.[63] The international development group Action Aid recently documented 132 cases of patenting of genes – including such staple crops as corn, potato, soybean, and wheat – that had evolved in developing countries.[64]

The benefits of owning the definition of plant and animal species and other life forms are not limited to agriculture. Pharmaceuticals companies have their eye on the potential windfall as well. Corporations seek out biological resources in impoverished countries, patent them, and then market their new commodity in wealthy countries using intellectual property protections to ensure their singular control of the profits.[65] For example, the pharmaceuticals giant Eli Lilly has developed two cancer drugs from the rosy periwinkle of Madagascar. Cancer patients have benefited greatly. Eli Lilly has made hundreds of millions of dollars. The people of Madagascar, on the other hand, have gained nothing.[66] Pharmaceuticals companies argue that such biological resources should be treated as common resources, so they owe no debt to Madagascar or its people. However, the government and people of Madagascar, and countless other communities around the world, point out that the resulting drugs are not treated as common resources. Quite the contrary, they are sharply controlled by intellectual property laws. So why should the biological resources of the poor not enjoy the same protections? Moreover, they point out, if we truly wish to preserve these biological resources and the ecosystems on which they depend, the communities that live there must be given some sort of stake in their protection. This will not happen if the community receives no benefit from their use.

Yet finding unidentified biological resources is only the most modest version of so-called biopiracy. Pharmaceuticals companies regularly identify traditional knowledge from healers in Africa and Asia, patent it in the West, and reap the profits of their monopoly. For example, the US National Cancer Institute recently discovered the value of the plant *Homolanthus acuminatus*. Traditional Samoan healers have for generations been grinding up the stem of this plant and steeping it in hot water as a treatment for viral infections. Nevertheless, it is Western pharmaceuticals companies, not the Samoan government, people, or village healers, who will benefit from the millions in future sales of the drug. In the highlands of Peru the high-altitude plant maca has been grown for centuries by indigenous peoples and has a reputation as a fertility enhancer. Despite its traditional use, a predatory US patent for the plant's 'natural viagra properties' was recently filed.[67] Similarly, the hoodia plant of the Kalahari desert, long used by bushmen on long journeys because it helps stave off hunger pangs, was recently patented for its potential as a diet drug. When confronted with charges of biopiracy at an international environmental summit in South Africa, representatives of the Pfizer pharmaceuticals company that will market the product quickly declared that this was merely a misunderstanding, 'not an exploitation story.'[68] Public attention helped ensure that the indigenous tribes will share in the profits. In most cases, however, corporations are not held to account. The cost of taking a multinational business to court is prohibitive for a poor country, so corporations enjoy a free hand in their theft.[69] And, in the end, yet another sacred gift of nature is made a cold commodity, bought and sold on the global market that drives us deeper and deeper into ecological disaster.

Water as a commodity

Ownership need not require invention, discovery, or even piracy. Sometimes just the claim of ownership, and the corporate power to back it up, is all that is needed. The corporate takeover of the world's water is a key example of this, and of the moral and environmental consequences of commodifying nature.

The world is running short of fresh drinking water. Although it may seem surprising, the earth has a limited supply. Only about 2.5 per cent of the water on the planet is freshwater, and two-thirds of that is permanent ice and snow. That leaves less than 1 per cent for us to share. And a large percentage of this is wasted or contaminated. Indeed, if we continue squandering water resources, in less than half a century we

5.10 / Rising global water demands (cubic kilometres per year)

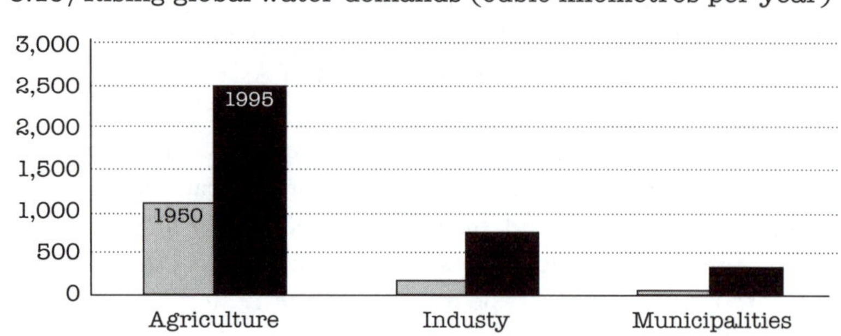

Source: *Crops and Drops: Making the Best Use of Water for Agriculture*, FAO, 2002.

will have wasted or contaminated enough water to cover the entire US, including Alaska, with more than five foot of water.[70] Once again, the perversity of narrow economic calculations leads to tremendous waste of a precious resource.

The situation is already in crisis in many parts of the world. Over a billion people lack access to clean water. Some 5 million people die every year from illnesses caused by drinking unsafe water, many of them children. Nearly 2 billion people live in regions of water scarcity, making food production and development difficult. And the future looks bleak. The UN's Environmental Program reports that the world's natural underground reservoirs are under increasing stress, threatening the source of water for over 2 billion people.[71] Thanks in large part to the increasing demands of industry and agriculture, water consumption has increased at twice the rate of population growth, and it continues to grow quickly.[72] Little wonder that the CIA has predicted that regional conflicts over water will be heightened by 2015.[73]

A combination of wasteful depletion, wrong-headed diversions, and pollution are leading to dramatic shortages in freshwater resources. Rivers and lakes are shrinking, groundwater sources are being depleted far faster than they can recover, and populations around the world are feeling the first effects of a looming crisis. After a global survey of water resources and usage, the World Meteorological Organization estimated that perhaps a third of the world's people will experience water shortages by 2025.[74]

The staggering global increase in large dam construction is one major problem. Since 1950 there has been a more than sevenfold increase in the number of large dams around the world.[75] The consequences of the more than 45,000 dams have been huge losses in forests and wildlife habitat;

5.11 / No access to water
(% without access to improved water source)

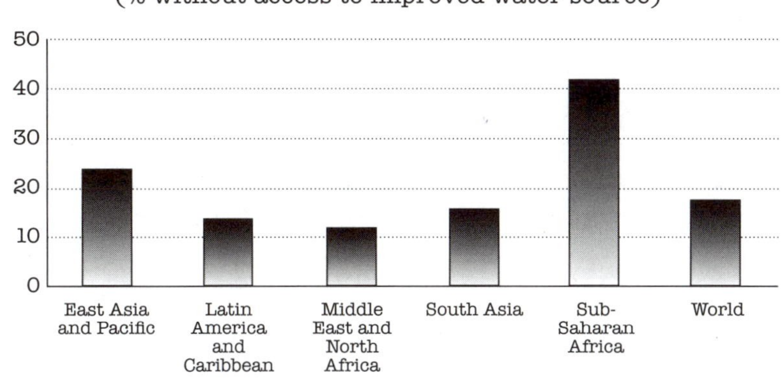

Source: World Bank, *World Development Indicators 2003*, Oxford University Press, New York, 2003.

the loss of aquatic biodiversity both upstream and downstream of the dam; the degradation of downstream floodplains, wetlands, and estuaries; and in too many cases an irreversible loss of species and ecosystems.[76] And, since these massive projects often prioritize the water needs of industries over the needs of people, access to safe freshwater for millions is also threatened.

The disruption of the water cycle can lead to unforeseen water shortages. Only 2 per cent of all US rivers and wetlands are still free-flowing and undeveloped. The US has already lost more than half of its original wetlands. This has placed large numbers of fish and amphibians at risk of extinction, and it has led to local water crises in some areas. Wetlands capture and store rainfall and snowmelt, and function to purify water that runs through rivers and aquifers. So they play a central role in the water cycle. When we destroy them, we compromise the delicate water cycle functions upon which all life depends.

The expansion of industrial agriculture has intensified this stress. Agriculture now accounts for about 70 per cent of global water use, and its overall use is increasing.[77] At the same time, as much as 75 per cent of irrigation water evaporates or is lost as runoff, never reaching the plants intended to be watered.[78] Agriculture not only claims great amounts of fresh water through irrigation; it places tremendous pressure on our remaining wetlands by dumping high levels of nitrogen, phosphorous, pesticides and sediment in surface and ground waters. Once again, inefficiency is rife in our use of water.

5.12 / Proportion of global population without safe water (%)

Asia	63
Africa	28
Latin America and Caribbean	7
Europe	2

Source: Rob Edwards, 'The Road from Rio', *New Scientist*, vol. 175, no. 2356, 2002.

The American Southwest provides a troubling example of the rising water stress in the US. Overexploitation, principally for agriculture, has been intensified by population growth and industrial expansion to define a serious threat to sustainability.[79] Rivers, lakes, and streams are drying up or are unable to support native species. Rural communities are finding their water access threatened by overdevelopment and are suffering social and economic decline.[80] Overall, the problem is quite simple: the gap between the amount of water being used and the natural rate of replenishment is growing dangerously large. In New Mexico's high desert, local communities lost their access to water when it was diverted for use by high-tech industries. In California, the battle between urban water demands, agricultural interests, and environmentalists has been raging for decades. As population growth outpaces water supplies, and as we continue to contaminate our natural waterways, this battle will intensify.

Natural bodies of water are feeling the strain all around the world. The famous Yellow river in China very often does not reach the sea for several months at a time, causing severe sedimentation of the riverbed, movement of saltwater up the riverbed, and crop losses in China's Shandong province. The coastal lagoons in the Nile delta have grown severely polluted, leading to serious threats to bird and fish populations. The Indus river delta in southern Pakistan is drying out and turning saline because of the increasing strain of irrigation. In the Rio Grande, endangered species are threatened by extensive water withdrawals. The Colorado, Missouri, and Mississippi ecosystems are highly vulnerable. River systems around the world are threatened by dropping water levels.

Most striking of all, the Aral sea is quite literally disappearing in what has been called the greatest man-made natural disaster in history. This Central Asian body of water was once the fourth largest lake in the world. However, heavy extraction for irrigation, mostly for cash crops for export, like cotton and rice, has literally drained the lake of life.

5.13 / Growing water contamination in the US (% of tested water found to be polluted)

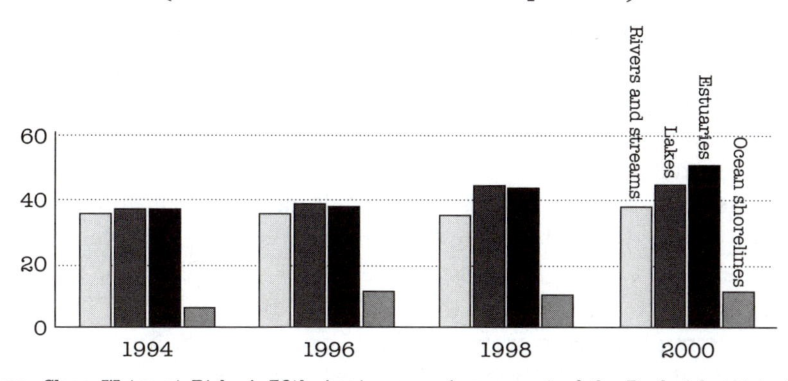

Source: *Clean Water at Risk: A 30th Anniversary Assessment of the Bush Administration's Rollback of Clean Water Protections*, Report of the Natural Resources Defense Council, Washington DC, 2002, www.nrdc.org; EPA, Office of Water, *National Water Quality Inventory*, 1996 and 2004, www.epa.gov.

Since 1960 the lake has shrunk to one-third of its previous area, a loss of 18,500 square miles. Salt has accumulated, picked up by prevailing winds and spread on deteriorating cropland to the south. Salt and dust storms are common. The entire delta ecosystem is threatened. Respiratory diseases, birth defects, cholera, cancer of the blood and other health effects are on the rise.[81]

This may be an extreme example or it may be the shape of things to come, particularly as the effects of global climate change intensify water shortages around the world. Recently a team of researchers from the Union of Concerned Scientists and the Ecological Society of America undertook the most comprehensive assessment of the impact of global warming on the Great Lakes region of the US. This two-year study utilized the most sophisticated models of the earth's climate available. In the end, they concluded that the Great Lakes region will look more like the American southwest by the end of the century. Water levels will decrease significantly and the general climate will grow arid.[82] As a result, water stress is likely to intensify.

In other parts of the world there is no need to wait for climate change to take its toll; countries in the arid or semi-arid regions of Africa and Asia are facing an immediate crisis. Rapid and unchecked industrial expansion has dramatically increased the rate of surface water pollution and the poisoning of groundwater.[83] Pesticides, heavy meals, and persistent organic pollutants contaminate water sources on which entire populations are dependent.[84] However, in many areas, the most pressing source of water pollution is human waste. Over 90 per cent of sewage is dumped untreated into waterways. As a result, every year, more than 2 million people, most of them children, die from sanitation-associated diseases.

Because of these growing stresses, populations around the world have grown increasingly dependent on water pumped from underground sources, generally called groundwater. Since about 97 per cent of the planet's fresh water is stored within the earth, this makes sense. The trouble is that, with massive increases in the use of agricultural chemicals, industries, and other waste, serious pollution of groundwater is being discovered almost everywhere. In California's Central Valley, for example, nitrite levels in groundwater have increased nearly threefold since the 1950s. The United States Geographical Survey found that, across the US, a quarter of the sites sampled were contaminated with two or more pesticides.[85] Unlike pollution of rivers and lakes, the contamination of these underground sources by farm, factory, and urban pollutants is essentially irreversible. The natural cycle of groundwater is extremely slow. Water filters through sand and gravel for hundreds of years. The average residence time for groundwater is 1,400 years.[86] It can therefore take thousands of years for an underground source to cleanse itself.

Furthermore, pollution of groundwater is extremely difficult to predict or localize. In many cases, we are just discovering the contamination caused by polluting practices that took place half a century ago. For example, although DDT was banned decades ago, this poison is still found in US groundwater. Contamination flows through rock and sand, spreading in ways that are difficult to forecast, seeping slowly but inexorably into groundwater.

The contamination of water sources increases the demands on the remaining freshwater. In the United States, China, India, and the Middle East, major aquifers are being drained at rates that far exceed their rate of natural recharge. This overdepletion can have a multiplying effect on the overall rate of loss by increasing the concentration of pollutants in the remaining groundwater. Overdrawing water can also cause sediments to compact, thereby shrinking the capacity of aquifers.

With scarcity comes the potential for profit, and this has not escaped the notice of wealthy corporate entrepreneurs. Global Water Corporation has noted the growing scarcity, declaring, that 'Water has moved from being an endless commodity that may be taken for granted to a rationed necessity that may be taken by force.'[87] The World Bank estimates a $1 trillion global market for water.[88] Monsanto, one of the major producers of genetically modified crops, is currently planning its entry into the global water market. In an often-repeated but revealing quotation Robert

Farley of Monsanto declared that 'what you are seeing is not just a consolidation of seed companies, it's really a consolidation of the entire food chain. Since water is as central to food production as seed is, and without water life is not possible, Monsanto is now trying to establish its control over water.'[89] Evidently, what is a humanitarian and ecological crisis to those who suffer from water shortages is a business opportunity for Monsanto. As a company strategy paper has it, 'we believe that discontinuities ... are likely, particularly in the area of water and we will be well-positioned via these businesses to profit even more significantly when these discontinuities occur.'[90]

Indeed, water is the new darling of global profit-mongers. As *Fortune* magazine noted, these companies are 'betting that H_2O will be to the 21st century what oil was to the 20th.'[91] Annual profits from the water industry, the magazine notes, are now about 40 per cent of the profits from the oil industry. Since only about 5 per cent of the world's water is owned by private interests, the potential for expansion is huge. Privatized water is already a $400 billion a year business. The two largest players alone, Vivendi Universal and Suez, control water services to over 200 million people in 150 countries.

This privatization threatens those who are most vulnerable to water shortages. Although public water suppliers in impoverished countries are often inefficient and run down, they have at least provided many poor urban communities with cheap or free water. While international corporate takeovers promise to improve services, they very often exclude the poor from an affordable clean water supply. Corporations aren't in it to provide better lives in poor urban and rural communities; they're in it to make a buck. The improvements they bring often include better services for the wealthy and more effective bill collection, but no improvement in water access or costs for the poor. In fact, costs usually jump up quickly, leaving many without an affordable source of safe water and reaping disastrous public health consequences.[92]

In one particularly disturbing instance, the World Bank pushed for the privatization of water services in Bolivia's third largest city, Cochabamba. In this semi-desert region, water is scarce. The sole bidder, Bechtel was granted a forty-year lease to the city's water system in 1999. Water bills quickly reached $20 per month – in a region where the minimum wage is $100 per month. Protests ensued, shutting down the city for several days. Within a month, millions of Bolivians marched to Cochabamba and staged a general shutdown of the city. The government promised to reverse the price hike, but instead later imposed martial law, arrested activists, and censored the media. The Bolivian government revoked the contract; and Bechtel, the company that had hoped to take over the water supply and reap substantial profits, is now pressuring Bolivian officials for compensation for their efforts and lost potential profits.

Cochabamba is not a lone example. Ten million residents have had their water cut off since the implementation of a water privatization program in South Africa. In Kwazulu–Natal nearly 100,000 contracted cholera when water services were cut off due to nonpayment.[93] In India as much as 25 per cent of total household income needs to be spent on water. Poor residents of Lima in Peru pay private water vendors as much as ten times the amount paid by wealthy city residents who are served by a municipal water system.[94]

Water shortages for the poor aren't the only cost of privatization. The corporate water industry, particularly bottled water, generates tremendous waste. Growing at 20 per cent annually, nearly 100 billion liters of bottled water are sold each year. To fill these bottles, corporations are buying up farmlands, indigenous lands, wilderness, and community water systems.[95] Massive pipelines and supertankers are constructed to carry water to their markets, bypassing thirsty populations surviving off inadequate or unsafe water, and adding uncounted environmental costs. The effect is perverse: while millions around the world go thirsty, massive amounts of energy and resources are used to process, bottle, and ship millions of tonnes of water thousands of miles around the globe. French water is sent to California, and Californian water is sent to France, and the ecological footprint of a precious natural resource skyrockets. The kicker is that these designer waters often only meet the same or lower standards than the available tap water.

There are even more disturbing examples. Global Water Corporation has forged an agreement with Sitka, Alaska, to export 18 billion gallons of glacier water each year. The water will be shipped for bottling in one of China's 'free trade zones' where weaker environmental laws and cheaper labor will facilitate greater profits.[96] Huge plastic bags are being constructed to transport massive quantities of water across the oceans. Wouldn't it be a lot easier simply to be more responsible with the water we have? Yes it would. But that wouldn't be nearly as profitable.

The increasing commodification of water means that it will not be protected as a sacred, life-giving resource. The solution to global scarcity and thirst will not be more equitable distribution, conservation efforts, or limits on corporate waste. Instead, in a system controlled by corporate profit seekers, the market will ensure the most profitable allocation of water, even if this risks the death of millions, the destruction of irreplaceable aquifers, rivers, and lakes, and mounting pressure on our other natural resources.

Governments had often argued in the past that access to such basic necessities of life as water or healthcare should not be impeded by trade agreements. However, with the mad rush to unregulated markets, water is being treated like any other commodity. The great perversity, of course, is that, as with food, we have the capacity to provide plenty of water for all the earth's people in a sustainable way. This doesn't require any scientific breakthroughs; we know how to do it now.[97] Reclamation systems, drip irrigation, conservation, and other significant changes in the way water is used and allocated could help provide safe and accessible water to every person on earth. This would, however, require that we prioritize the earth and its people over profits. The opposite is taking place. Corporations would like to determine the future of these precious resources, and water will go to those who can best afford to pay and bypass those who cannot. We need to stop this plan dead in its tracks.

Changing the rules / 6

I know what you're thinking. Why aren't we doing anything? We're on the brink of social and environmental devastation. Heck, we're not on the *brink* – we're right smack dab in the middle of the devastation, and we don't seem to be changing course. Where's the 'environmental president'? What are our lawmakers doing? Where the hell is democracy? Corporations do not run the country, we do... or at least our elected representatives do... or, they're supposed to. It's their job to deal with things like this, to ensure our safety and security and to preserve our common interests. What are they up to? Why has it come to this?

Well, it would be easy to blame a particular president or industry. And it's pretty clear that certain policymakers and corporate leaders have shown a blatant and catastrophic disregard for the environment and public welfare. However, George W. Bush did not destroy the global environment single-handedly. Monsanto did not singularly and maniacally engineer global hunger. The problem is deeper and more pervasive than anything for which one politician or corporation could possibly be responsible.

It's certainly true that over the past few decades we've taken some steps in the right direction. The US has an extensive network of environmental laws, policies, and practices that protect the environment – or at least that's what they're supposed to do. We've defined a collection of rules and regulations that count everything from how much toxin we dump into rivers to how many animals are at risk of extinction; and when the numbers get really bad, we sometimes do something about it... *sometimes.* But in the end, these only amount to minor adjustments to the machine. In the best cases, we penalize a particularly destructive corporation enough to make it stop, or at least reduce its most caustic practices. Yet even this is often just a political sedative, designed to assure the public that everything is OK while very little or nothing changes. In the big picture, business as usual persists. Even those policies that do seem to offer real change are only meaningful if they're enforced, and virtually none is carried out as intended. So, despite the environmental regulations that have been adopted with great fanfare and self-congratulatory zeal on the part of lawmakers and environmentalists alike over the past three decades, things continue to get worse. Maybe we've slowed down the rate a bit, and that's a good thing; but the trend is still definitely and clearly for the worse.

The truth is, while explicitly cast as protections for the fragilities of nature, environmental regulations really protect the status quo. Our environmental rules have two faces. One promises safety and environmental protection to the people. The other delivers a continued free hand to corporations (and a pacified public to policymakers). Thus, while we'd like to think that we have policies in place that protect us and ensure the defense of our oceans, rivers, air, and forests, the truth is that none of these is protected with the same vigor with which we protect the interests of powerful industries.

Cleaning up our act?

Environmental protections in the United States emerged slowly, mostly as piecemeal responses to public outcry, and of course always moderated by the influence of industry. For example, the early efforts of preservationists like John Muir around the turn of the century won the creation of a national park and forest system; but in the end this system served the interest of corporations first by ensuring cheap raw materials for timber and mining industries.[1] The federal government didn't protect forests; it managed them as crops to be harvested for profit by private industry. Similarly, when Rachael Carson published her inspired call for the protection of nature, *Silent Spring* (1963), legislators facing popular outrage acted to restrict the use of DDT. But the sale of DDT to impoverished countries around the world was still allowed; and other chemicals that were just as environmentally devastating but not yet known to most citizens were still not regulated in the US. In these two examples and countless others, policymakers did what they had to do to appease public pressure, but they were always careful to cater to corporate interests first.

By the 1960s, the environmental crisis in the US had become increasingly evident. The state of the nation's rivers and lakes, for example, raised serious alarm. Lake Erie had been declared dead. Major rivers like the Hudson and Mississippi were treated as open sewers. The most dramatic example was provided on 22 June 1969, when the Cuyahoga river in Cleveland burst into flames – or, rather, the oil and other industrial waste covering the river burst into flames. The blaze reached the equivalent of several storeys into the air. This had happened before, and was considered unspectacular by the locals. But, with the help of an article in *Time* magazine, this time the river fire gained national attention and made the sorry state of the nation's rivers clear to a generation of Americans. A massive oil spill off the coast of Santa Barbara, California, on the afternoon of 29 January 1969 sent a similar message. A blowout on an offshore oil rig allowed 200,000 gallons of crude oil to bubble to the surface. An 800-square-mile slick coated the shoreline with thick and deadly tar for dozens of miles. Seal and dolphin corpses were washed up on the shore. Thousands of sea birds were killed. The destructive consequences of unrestrained industry were growing hard to ignore.

By the early 1970s, a new generation of researchers and advocates were joining Rachel Carson's call for reform. Books like Paul Ehrlich's *The Population Bomb* (1968), Charles Reich's *The Greening of America* (1970), and Barry Commoner's *The Closing Circle* (1971) called attention to an increasing ecological crisis. Demands for real reforms grew louder. A growing population of middle-class Americans began to take note of the threat to their families' health. In 1970, on campuses and in communities around the country, the first Earth Day defined a day of celebration of nature's beauty and a day of education on the growing emergency. Some charged that the effort was unduly focused on middle-class suburbs and campuses, ignoring the concerns of labor, the poor, and communities of color. A few anti-war activists accused Earth Day organizers of distracting attention from the anti-war struggle and playing into the hands of politicians who wanted to take some momentum away from the anti-war movement just days before the bombing of Cambodia began. These debates would mark enduring strains within the environmental movement.

The piecemeal and pitiful approach that would define the next three decades of environmental policy was evident even at the formation of the Environmental Protection Agency (EPA) in 1970. President Nixon's formation of the EPA was hardly an unambiguous victory for Mother Nature. Nixon's interest in environmental issues was opportunistic, a defensive ploy against his political challengers and desngned to ensure that whatever protections did take form would not threaten core industrial interests.[2] With public outcry growing, his Democratic rival Senator Edmund Muskie proposed landmark environmental legislation that clearly went further than corporate leaders or the Nixon administration wanted. Realizing that some environmental reforms were politically inevitable, industry leaders moved to shape legislation to ensure that the impact of the new rules on their bottom line would be minimal.[3] Nixon, wanting to please his corporate supporters by pre-empting the more extensive legislation of his rival while simultaneously hoping to gain credit with green voters, moved alternative, weakened, and unenforceable legislation.[4] When his Democratic rival's bill gained bipartisan support, Nixon worked to weaken it. And when it passed, he took the credit. So, in the end, this marked a victory for environmentalists, but clearly a tempered victory. Subsequently, the lack of commitment to enforce these new laws and the woefully inadequate funding allocated to the task tempered the victory still further.

This is a common pattern. Corporate leaders and their policymaking allies claim green credentials by championing watered-down, ineffective but fine-sounding environmental legislation. When public demands can no longer be ignored, industry leaders make sure that they have a strong hand in the design of the resulting policies. In the case of the EPA, they championed feeble federal intervention in the name of what they called 'regulatory consistency.' That is, they didn't want to deal with different environmental laws in each state and county, and they needed to stave off the risk that some local governments might enact laws that restrict their actions too severely. With their allies in Washington, business leaders have almost always been able to ensure an outcome in their favor.[5] The result has been halfway measures that, while an improvement on the hands-off policy that preceded them, were hopelessly inadequate to protect against further destruction, much less repair the devastation of the past. When public pressure is so great that the legislation cannot be watered down, corporate-friendly policymakers have ensured that the law is subsequently not enforced too rigorously, or they have managed to have funding for enforcement sliced once public attention wanes.

Efforts to protect US waterways offer a sharp example of this pattern. Despite President Nixon's veto, public outcry catalyzed by events like the Cuyahoga river fire ensured that the Clean Water Act took full effect by the late 1970s. Its goals were clearly stated: by the mid-1980s, all discharge of pollutants was to be completely eliminated from navigable waters, water quality in fishable and swimable waters was to be improved, and all pollutants in toxic amounts were to be eliminated and prohibited in any of the nation's waterways. Pollution from 'point sources' – factories, treatment plants, and the like – were to be sharply regulated. States were required to set water quality standards that met with federal requirements. If they didn't, the federal government had the authority to step in. This federal primacy was indispensable to stop what some called the 'Mississippi syndrome,' a dynamic that we now see internationally as the 'race to the bottom.' Before federal oversights, states that wanted to attract industries felt pressure to reduce their barriers to environmental pollution. In effect, local and state leaders would bargain away their clean rivers, lakes, and forests to attract polluting industries. Moreover, if one state did take the hard steps needed to improve its water quality, discharge from other states could still pollute their rivers, streams, and lakes. Environmentalists argued that only federally enforced minimum standards, as the new law required, would stop this downward spiral.

Sounds promising, right? Well, even moderate gains such as these sparked an anti-green backlash when the Reagan administration took office in 1980. Federal policy shifted toward a messianic fixation on free markets and corporate-friendly reforms. Reagan's promise to get 'government off the people's back' – populist rhetoric meaning to eradicate government protections for the environment, social welfare, and public health – meant giving a free hand to corporations. Administrators were clearly and unabashedly contemptuous of environmental protections. James Watt, a pro-development lawyer who made a career of bringing lawsuits to defeat environmental protections, was made secretary of the interior. Watt's vision of environmental stewardship was made clear when testifying before Congress; asked if he believed that natural resources should be protected for future generations, he answered, 'I do not know how many future generations we can count on before the Lord returns.'[6] Anne Burford, a corporate attorney with antagonism toward any government regulation, was given the keys to the EPA. As one EPA memo had it, businesses were 'the primary constituents of this administration.'[7] Cost–benefit analysis was made mandatory for any environmental or public health policy. The government's capacity to monitor and reduce environmental damage was gutted. Severe budget cuts meant that even the day-to-day efforts of the government to ensure public health were crippled. In one sudden jolt, two decades of hard-won environmental protections were brought to a standstill.

So where did that leave the nation's water? Federal funding for clean water projects all but disappeared. State enforcement of environmental protections waned seriously. Industries were left to monitor themselves. If not for the pressure of active citizen groups and environmental advocates, there would have been no meaning at all to the Clean Water Act. As it was, with insufficient enforcement and woefully inadequate funding, the Act was clearly more of a symbolic palliative to public concerns than a real attempt to protect public health and the environment.[8] The public were given a law which they were led to believe would fix the problem, and the corporate order could go on as before, largely unaffected by the anemic new rules.

By the end of the 1980s, it was clear that these setbacks weren't just about the Reagan administration. A pattern of weak enforcement, minimal funding, and general corporate dominance of public decision-making on the environment had become the norm. The first Bush administration's Council on Competitiveness used the premiss of protecting America's corporate dominance to review and strike down any regulation or policy that might impair industry's continued profits. Under the leadership of vice-president Dan Quayle, the Council's first act was to order the EPA to cancel a proposed recycling program on the grounds that it would put an undue economic burden on the owners of municipal incinerators.[9] Goodness forbid we burden the industries that regularly pump hundreds of tonnes of deadly toxins into the air.

Although the Clinton administration offered an improvement on its predecessors, the fundamental pattern was not greatly changed. Wanting to cast himself as the environmental president, Clinton brought in selected environmentalists to key posts. Bruce Babbitt, a former governor of Arizona and head of the League of Conservation Voters, was put in charge of the Department of the Interior. The Council on Competitiveness was replaced with a White House Office on Environmental Policy, and George Frampton, former head of the Wilderness Society, was selected to lead it. It seemed to many environmental leaders who had fought outside the corridors of power for so long that they had been given the keys to the kingdom. But they soon realized that their room for maneuver was desperately small. Corporate interests and economic performance (measured perversely as growth in GNP and consumer spending) were the criteria that drove policymaking. And despite Clinton's promise to balance environmental and economic interests, he proved much more inclined to favor the latter. The use of 'market-based instruments' and cost-effective analysis grew, ensuring that only the most restrained environmental protection would be adopted.[10]

Nevertheless, having environmentalists inside the administration helped. Environmental targets were strengthened. A major assault on environmental protections from conservative Congress members was staved off. Natural resource policy was made more sensitive to environmental concerns. But the overall pattern changed very little. In his first Earth Day address, Clinton promised to stabilize greenhouse gas emissions at 1990 levels by the year 2000.[11] Yet when he left office seven years later, America's emissions were 15 per cent higher than 1990 levels. The Clinton administration pushed NAFTA despite the clear environmental risks. Clinton's 'compromise' with the logging industry facilitated continued destruction of old growth forests and gave a free hand to logging on national land.[12] After serious backsliding, the Clinton team ensured that massive federal subsidies to mining industries were continued. In general, although the balance had been altered somewhat, Clinton's policies never directly addressed the broad structures that drive our ecological suicide. A neo-conservative Republican takeover of Congress two years into Clinton's first term cemented business's hold on federal policymaking. Clinton fiddled with the controls a bit, but kept the machine running.

Whenever efforts at real change were attempted, they were quickly ruled out by a concerted pro-industry phalanx of policymakers, lobbyists, and corporate money. In 1993, for example, when newly elected President Clinton proposed the so-called BTU tax, a tax on all forms of energy (measured in British Thermal Units, hence the name), the opposition was sharp. Environmentalists preferred a tax on fossil fuels; but, knowing this would be a political nonstarter, since it would face sharp opposition from coal-producing and coal-using industries, Clinton proposed a tax on all energy sources. The stated aim was not to protect the environment but to reduce the budget deficit. Still, the tax would have raised the cost of gasoline and electricity moderately, and any additional cost to energy consumption would have a favorable impact on emissions. Environmentalists hailed the idea as a way of encouraging conservation. It didn't take long for lawmakers from energy-producing states, reinforced by industry lobbyists, to kill the idea definitively. Of course, tax programs in general are politically easy to oppose. Environmentalists in the Clinton White House received the signal loud and clear: Real change, the kind of change they had hoped for, the kind that was necessary, the kind that could save us, was not on the cards. An environmentally meaningless 4.3 cent per gallon tax on gasoline was put in place instead; and, as oil prices dipped and the popularity of SUVs rose, Americans were consuming more fossil fuels and producing more carbon dioxide than ever. To assure the fossil fuel industry that he was their friend, Clinton opened the Strategic Petroleum Reserves twice to keep gas prices even lower than the market would have allowed. Placing any constraints or additional costs on America's rabid consumption seems to require greater courage than any recent president has been willing to display.

Clean water and air

Where does that leave us? Well, it leaves us with little hope that the kind of structural change that is needed to save us from ecological disaster will emerge from the established power brokers. But what about small changes? What about the regulations and practices that are meant simply to protect human health or reduce the most destructive impacts of profit mongering? Let's get back to the Clean Water Act and the nation's waterways. Since the 1970s there have been some improvements. New sewage treatment plants have been built with the support of federal funding and corporations were forced to clean up their acts at least a little. But, as I mentioned, the Act has unfortunately fallen far short of its initial promise. In fact, 'noncompliance' – that is to say, violating or ignoring the law and polluting in excess of legal limits – is commonplace. And authorities do very little to stop such violations and even less to punish them. The EPA's own internal study found that about a quarter of the country's major industrial plants and water treatment facilities are in serious violation of pollution standards at any one time. Many exceed pollution limits for toxic substances by over 100 per cent. Some 13 per cent of industries were found to exceed legal limits for toxins by over 1,000 per cent! What do federal and state authorities do to protect us and punish these clear and egregious violations of the law? In 85 per cent of the cases, the EPA did nothing. That's right, not a thing. In the 15 per cent of cases where formal action did take place, fewer than half of the violators actually paid a fine. And the average fine levied was about $6,000.[13] The EPA's own audit found the agency's enforcement system 'incomplete, inaccurate and obsolete.'[14] And when faced with its own audit detailing its failures, the EPA Office of the Inspector General reported that 'the Office was reluctant to change its current way of conducting business. However, the current way of conducting business was marginally effective.'[15] Little wonder that about 40 per cent of the nation's waters do not meet safety standards.[16]

A general practice of assumed non-compliance has been defined. Industries have evidently found that it's cheaper to pay fines than install needed pollution-abatement equipment; and policymakers are hesitant to irritate business by raising the penalties. Deadlines are regularly and generously extended at the discretion of administrators. Enforcement is pursued with dramatic inconsistency from state to state, with the amount industries are allowed to overpollute and the diligence with which polluters are stopped varying widely.[17] Only ten states had a compliance rate above 90 per cent. In almost half the states, a quarter or more of major industries or plants were found to be violating environmental laws repeatedly.[18] As a result, tonnes of toxins are still being released into our rivers, lakes, and streams. Mercury poisoning has made it unsafe to consume fish from many American waterways. There are tens of thousands of releases of raw sewage – bacteria, fecal matter, industrial wastes, viruses, and other contaminants – into waterways every year. The EPA estimates that as many as 3.5 million Americans get sick every year from swimming in water contaminated with sewage. The real kicker is that some of the largest sources of pollution go untracked; we don't really know how bad they are, and we have no effective rules to deal with them.

This pattern isn't unique to the Clean Water Act; it's evident in the policies that pretend to protect our air, endangered species, forests, food, and other precious things. For example, in 1995, under pressure from the new Republican-controlled Congress, the Clinton EPA announced a policy of regulatory self-audits for polluting industries. That's right, a corporation would be allowed to determine for itself whether it had violated environmental laws. If it had and it reported them, its penalties would be reduced or eliminated. Of course, if it didn't report them, no one would know anyway. Proponents of the policy argued that perhaps if industries did not fear penalties or criminal prosecution they would have an incentive to admit and fix problems. The more reasonable among us can see this as a clear giveaway to industry. They can knowingly violate environmental laws for years, place communities in jeopardy from their waste, destroy precious habitats, contaminate rivers, and enjoy the windfall profits that are possible when pollution control requirements are ignored. In the end, all they need to do is admit to the mess and they get off without penalties. And, since the EPA's budget and political support are far too feeble to ensure proper regulatory enforcement (hence the need for self-audits), even if they didn't admit to their malfeasance, they were still unlikely to face penalties.

The policies that are supposed to protect our air tell an important story about compromise. Like the Clean Water Act, the 1970 Clean Air Act (CAA) seemed to offer reason for optimism. The CAA was supposed to empower the federal government with broad discretion to regulate industrial pollutants. It also called for a major reduction in automobile emissions. But once again it became clear very early on that half-hearted enforcement and capitulation to corporate interests would cripple this law. While the federal government could, and was supposed to, limit dangerous emissions of toxic chemicals, in reality regulators did relatively little. Only seven of the hundreds of harmful pollutants that are pumped into our air every day were regulated. Congress made revisions in 1977 that were supposed to fix the inadequacies of the law; but while the emission of lead and particulate matter did decrease, other toxic emissions continued to increase; and lethal quantities of many pollutants were still being regularly released.

In 1980, a ten-year study mandated by Congress examined the effectiveness of the Clean Air Act. Like the Clean Water Act, noncompliance had become the norm. A full 87 per cent of integrated iron and steel plants were out of compliance. Over half the country's smelters were not in compliance. Even the less polluting industries were only in compliance about 8 per cent of the time.[19] And most major cities exceeded federal limits several days each year.

The most disturbing shortcoming was the continued acidification of rain. The report's conclusions were clear: the environmental consequences of acidification were wide-ranging and serious. Previous attempts by Congress to adopt greater protections had been consistently beaten back by a coalition of utilities, the automotive industry, oil companies, and their congressional allies. However, by the end of the decade, public concern was strong enough to require some form of action. The first President Bush tried to play both ends, supporting a bill in public but then working behind the scenes to weaken it. Along with leading senators, Bush threatened a filibuster if key provisions of the law were not cut. Industry lobbyists, such as the American Petroleum Institute, the Chemical Manufacturers' Association, and the National Coal Association, worked closely with the Bush team to ensure that the resulting legislation was in their favor. Most notably, the so-called Clean Air Working Group, an alliance of various business interests, began an aggressive and well-funded effort to weaken any attempt at serious air pollution reductions.[20] In the end, White House and Senate negotiators went behind closed doors to hammer out a compromise.

On the face of it, the resulting 1990 amendments to the Clean Air Act seemed to offer the public reason for optimism. Power plants were required to reduce their sulfur dioxide emissions by 50 per cent of 1980 levels. Reforms were proposed to prevent smog, achieve national air quality standards, and establish exhaust and fuel standards. However, the closed-door sessions had clearly worked in favor of business preferences. The new rules took absolutely no direct steps to reduce greenhouse gas emissions. Utilities remained exempt from regulations on 189 ultra-toxic substances, while the nation continued to experience widespread violation of minimum safe standards for air quality. The Act did not address mercury pollution from coal combustion, despite the fact that essential food supplies were increasingly contaminated with dangerously high mercury levels. Even on the issues the Act did address, full implementation was expected to take up to twenty years and these deadlines could easily be extended. Even if the reductions on sulfur dioxide emissions were reached, scientists warned that this would not be nearly sufficient to prevent acid rain. More than this, the carefully crafted ambiguity of the regulations made enforcement questionable. (This was made evident in subsequent court decisions that struck down EPA regulatory actions.) Industry lobbyists must have been pleased. Environmental regulations had been standardized and rationalized under a single regulatory regime, thus avoiding the threat of tougher standards from states. Environmentalists had been largely barred from the negotiations. And nothing in the resulting regulations required a significant change to business as usual.[21]

The most controversial portion of the plan was meant to address acid rain. Industry had ruled out any sort of hard ceiling on acid-rain-causing emissions. A proposed direct tax on these emissions, which clearly would have most directly internalized the cost of acid rain and therefore led to reductions, was crushed by lobbyists for power plants and the coal industry. Instead a market for tradable pollution allowances was put in place. In this 'cap and trade' system, a 10 million tonne cap on sulfur dioxide emissions was established and industries were allotted a certain level of pollution allowances based on the plant's past power production. Corporations could then buy and sell excess allowances to meet their needs. So what you have in the end is a free market in toxic emissions. This approach, advocates argued, allowed for flexibility in how reductions were achieved. Companies that could realize reductions most efficiently would do so and sell their excess emissions allowances to heavy polluters that could not easily reduce their emissions.

There is, of course, an evident moral dilemma in the free trade of deadly toxic emissions. As Peter Bahouth, former executive director of Greenpeace, had it, 'If you were trying to handle drug problems in your community, you wouldn't be saying: "Let's try to work this out with the drug dealers."'[22] Ask yourself, would a policy that allows drug dealers to buy and sell the right to push drugs on school children be okay as long as it results in some reduction in drug sales? I hope your answer is no. Corporations were given a right to pump deadly toxins into the air. More than that, they could actually sell these rights to other polluters when they had extra! The air you and I breathe was effectively sold off to the highest bidder. No surprise, the practice was immediately popular among polluters.

This market-based policy had several other problems worth noting. First, it essentially rewarded big polluters. With renewable energy sources

6.1 / Acid rain emissions (million tonnes)

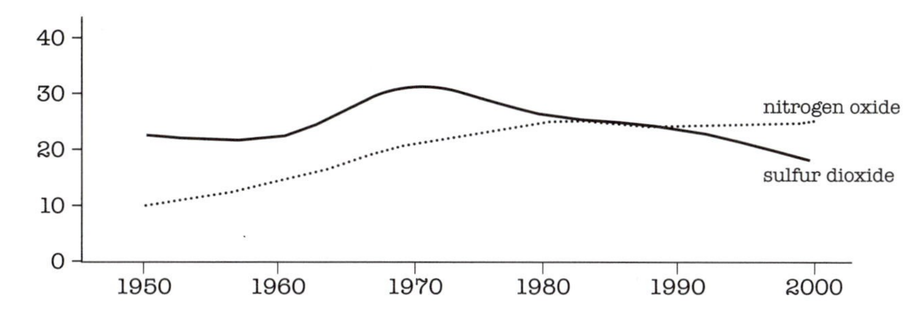

Source: US Environmental Protection Agency, Office of Air Quality Planning and Standards, Annual National Air Quality and Emissions Trends Reports, www.epa.gov.

excluded from the distribution of valuable credits, the biggest past polluters were granted the most valuable allowances. The worst polluters wouldn't need to reduce their emissions at all; they could simply buy allowances to continue poisoning the air legally. Second, such a system can create hot spots where highly polluting industries simply buy allowances rather than adopt cleaner technologies. Since these heavily polluted areas are most likely to occur near impoverished and marginalized communities, we once again end up with our most vulnerable communities exposed to dangerous pollutants. And, by handing their power over to a market mechanism, authorities have no means to balance the scales.

Emissions did go down, but not because of the trade in pollution credits. The plan was to reduce sulfur dioxide emissions by 10 million tonnes per year and nitrogen oxides by 2 million tonnes per year by the start of 2000. Initial targets were not only met but exceeded. Industry advocates for the cap and trade program argued that this proved the effectiveness of market-based pollution controls. But in fact relatively few trades between industries occurred. Only one company actually purchased allowances to meet the 1995 deadline, for example.[23] The greatest amount of the reduction came when coal plants in the Ohio Valley switched to low-sulfur coal. But the Clean Air Act didn't drive this shift.[24] Instead, the reduced cost of low-sulfur coal, long used in the West, and a sharp drop in transportation costs made low-sulfur coal a bargain for plants in the Midwest. Greater profits, not a desire to save the atmosphere, federal regulations, or even the market for pollution credits, drove the switch to low-sulfur coal. The kicker is that a switch to low-sulfur coal had long been advocated by environmentalists and rejected by industry leaders, who claimed it would be too costly. Since emissions allocations can be banked for future use, it's difficult to say what the long-term results of

6.2 / Percentage of major US pollutants from coal

Sulfur dioxide	60
Carbon dioxide	31
Mercury	32

Source: Now, PBS, at www.pbs.org/now/science/cial.html.

this market-based system will be. Emissions might rise significantly in the future and not violate federal requirements.

A recent and important research program by leading scientists in the field clearly indicates that acid rain has created far more environmental damage than had been predicted in the 1990s. Acid rain has, according to this research, 'set off a cascade of adverse ecological effects.'[25] The sulfur content of acid rain has declined modestly but remains very high compared to natural levels. Nitrogen oxide emissions have not declined and have in fact increased in some areas. Trees have not recovered as expected. The report concludes that the Clean Air Act is not likely to bring about recovery of the ecosystem.

The overall picture is all too clear. Regulators have simply grown too close to industry, and policymakers are simply too reliant on industry's support. This all-too-cozy relationship with industry has been evident across the board and is absolutely blatant in the enforcement of automobile emission standards, for example. An extensive investigation by the Commerce Committee of the House of Representatives found 'a pattern of gross negligence and striking indifference by the EPA' in the regulation of auto emissions.[26] It had been clear for some time that diesel truck engines in particular were emitting pollutants far in excess of federal standards, yet the EPA did nothing. Officials rubber-stamped the automobile industry's self-certification, allowing millions of tonnes of excess deadly emissions every year. The investigation found that the EPA was 'too close to the industry it was charged with regulating to adequately protect human health and the environment.' The corporate average fuel economy (CAFE) standards that now define federal emissions requirements for passenger vehicles are filled with loopholes. Minivans, SUVs, and pick-ups are allowed more lenient efficiency requirements than passenger cars, despite the fact that they make up nearly half the market in new vehicles. Cars like the Chrysler PT Cruiser get around emissions requirements by having a removable back seat, thus qualifying as a truck. Large vehicles, like the environmentally caustic Hummer, need meet no efficiency requirements at all.[27] Any proposal to close these loopholes, discuss further emissions reductions, or support the use of alternative fuels face powerful industry opposition. Legislators beholden to the automobile and fossil-fuel industries can be counted on to argue against even the most watered-down reforms. Hence, in the end, we can count on industry's anti-regulatory priorities to prevail.

Lowering our sights

The general pattern is repeated over and over. First, public concern leads to the enactment of environmental protections. The legislation is hailed by lawmakers as a real breakthrough, declared a genuine change in business as usual, and advertised as a sign of their clear commitment to the protection of the earth and public health. There is enough backslapping and grandstanding to fill the evening news. Middle-class environmentalists breathe a happy sigh of relief as they are assured that Mother Nature is no longer threatened. Of course, some environmental groups might protest, pointing out the inadequacies of the new rules, but they'll largely be ignored. Other environmental groups, those that deeply compromised their principles to endorse this moderate reform, will declare victory and send letters to all their members praising their triumph and asking for more support. However, the entrenched and well-funded efforts of corporations are also at play, in a much less public way. While the resulting rules offer broad promises and sweeping language, they are crafted to be strategically ambiguous, tempered, and carefully worded to ensure that key interests are not threatened. Subsequent lack of funding and the failure of enforcement ensures that these rules will never really challenge industry's continued pursuit of ever more profit at the expense of the environment. In the end, a green veneer carefully applied by PR firms and spin doctors allows lawmakers and businesses to assert their green credentials without really changing anything that matters.

The most disturbing fact is that these deeply compromised policies have too often been enacted with the clear support of those who we count on to be nature's great defenders. For example, the very concept of pollution trading and the implicit right to pollute entailed in these schemes were invented by policy specialists from the Environmental Defense Fund (now Environmental Defense). Evidently, in the face of entrenched corporate resistance, major environmental organizations have found it necessary to lower their sights and seriously soften their commitment to real change. After facing the Reagan anti-environmental backlash, organizations such as the Sierra Club, Environmental Defense, League of Conservation Voters, and others that had come to be dubbed the Group of Ten grew increasingly willing to compromise on their fundamental commitments in the 1980s. For the sake of marginal results that they could then claim as victories, they abandoned their deepest obligations. More and more, they talked about the use of market mechanisms to protect the environment and the requirement to balance environmental protections with the needs of industry. It's not simply that environmental groups were willing to compromise when necessary. Certainly, sometimes when a battle can't be won the best route is to find the middle ground and declare a partial victory. But environmental groups increasingly entered the fight with the *aim* of compromise, ready to give big business what they want in return for marginal concessions that allowed them to save face with supporters.[28]

This shift fundamentally undercut any hope for real change emerging from the major environmental organizations. People who study politics point out that winning a political argument is one form of power; but they point out that a more persuasive mark of power is the ability to define what the argument is about. If you get to set the terms and limits of the debate, you've got a good chance of winning it, and indeed subsequent arguments. The environmentalists who rallied for the first Earth Day and demanded fundamental change through the 1970s were proposing a new way of defining our place in the world. They were insisting on a basic shift in the political debate and challenging the fundamental values and principles of the system. By the late 1980s, the now established national environmental organizations had largely abandoned this effort and accepted the terms of the profit-centered, market-based corporate order. As a result, they were, and are, left fighting a rearguard battle. The terms and limits of the debate are sharply limited in favor of the status quo; and the best that many national environmental groups feel they can do is achieve moderate reforms within this system.

If environmentalists grew more submissive in the 1990s, anti-environmental forces grew more organized and powerful. A rural property owners' reaction to environmental policies had been around in the West since the 1970s. The so-called Sagebrush Rebellion opposed the federal government's expanded role in land and natural resources management that came with the environmental regulations of the 1970s. Hoping to take over valuable grazing lands and mineral rights on the cheap, these 'rebels' argued for states' rights and called for the remaining public lands under federal control to be transferred to state authorities and then privatized. Although this populist ploy for cheap land ended with little more than a whimper after Watt resigned in disgrace and the Reagan administration was forced to back away from its plans to sell off public lands, it blazed the trail for more effective anti-environmental populism in the 1990s.

With careful fostering and organization from corporate agents, the so-called wise-use movement emerged in the 1980s and 1990s as a loose alliance of logging, mining, agricultural and other extractive industries, combined with conservative interest groups. Rural communities that shared a distrust of federal policies, large landowners, and ranchers that worried about the impact of environmental protections on their pocketbooks were sympathetic to the cause; but the real foot soldiers of the movement were timber and mining employees and their families who had been convinced that environmental regulations threatened their livelihood. With a mix of grassroots appeal and corporate sponsorship, the environmental opposition called themselves outsiders and cast environmental organizations as the powerful, elitist, rich and self-serving establishment. The characterization was coarse, but it won them growing support among hunters, labor organizations, and middle-class Americans. The environmental movement seemed to be growing stale and now faced a coordinated opposition that went well beyond corporate lobbyists and their paid legislators.[29]

The showdown came in the Pacific Northwest, when environmentalists tried to stop the logging industry from destroying the country's last old-growth forests. Environmentalists attempted to use the Endangered Species Act to end the logging since these ancient forests were the only habitat in which the endangered northern spotted owl could survive. Their concern was not narrowly with the owl's fate alone; rather the owl is an indicator species – its demise would signal the deteriorating health of the entire ecosystem. Saving it would require that at least a million acres of old-growth forest be protected. The Reagan administration attempted to assist the timber industry by refusing to list the owl as an endangered species, pressuring federal agencies to back away from enforcement efforts, and in the end even attacking the Endangered Species Act itself as unnecessary and too strict. When courts ruled in favor of the environmentalists, the timber industry launched a full-scale public relations assault, accusing environmentalists of putting trees ahead of jobs. Later, Bush administration officials worked hard to undercut the intent of the Endangered Species Act; but it was Clinton who would enact a 'compromise' solution that left nearly half the remaining old-growth forest unprotected and set aside environmental laws to open most of the remaining old growth and vast tracts of national forest to logging of some sort.

The hypocrisy of the timber industry's claim should have been obvious. The previous decade of labor conflict had made it clear that industry leaders had no interest in preserving timber jobs or the welfare of their workers. Timber companies had shifted toward exporting raw timber to Asia in the 1970s, dramatically reducing the number of jobs linked to the industry but ensuring their continued profits. Federal subsidies to the tune of $100 million a year for private log exports sweetened the deal.[30] In addition, automation of the industry had led to fewer and fewer jobs at ever-lower wages. Louisiana-Pacific had already rolled back wages by 10 per cent and installed a two-tier wage structure over the protests of the union. Union-busting tactics by Weyerhaeuser allowed for dramatic wage cuts there as well. Labor unions resisted, strikes took place, but in the end the negotiating position of timber workers grew more perilous as wage competition from the growing Southern lumber industry and Canada increased the pressure. Of course, the sort of unrestricted cutting that the industry was doing would not ensure any job for very long.

Nevertheless, when environmental restrictions seemed to threaten their continued growth in profits, industry leaders and their friends in government quickly defined their opposition to environmental policies as a matter of saving jobs – the same jobs they had been eradicating for a decade. Environmentalists argued forcibly that old-growth forests must be preserved; they appealed to esthetics, morality, the precious gifts of unmolested nature. But they did so in opposition not only to the timber industry leaders but also to the workers whose livelihoods depended on timber and who had been convinced by their bosses that environmentalists were their enemy. Once again, environmentalists allowed the timber industry to define the lines of the debate. Middle-class environmentalists and rural timber workers ended up in a sharp conflict orchestrated by industry leaders, despite the fact that they had good objective reasons to be on the same side of the argument – both having a stake in a fundamental change in the industry toward sustainability. Environmentalists facilitated the task of industry with a condescending attitude that dismissed timber workers as unenlightened rednecks. For their part, bumper stickers declaring 'I love spotted owls – fried' revealed the depth of the timber workers' anger toward environmentalism. Radical environmentalists, frustrated with the deeply compromised policies of the mainstream organizations, staged protests, tree sits and, in a few cases, hammered metal spikes into trees to make them difficult to cut with chainsaws. One such spike was blamed for an accidental death. This inspired attacks against environmentalists as 'ecoterrorists,' radical tree worshippers who were callous to the concerns of people.

This is only one example of corporate efforts to blur the lines, to make victims into villains and villains into heroes. Knowing that most Americans consider themselves environmentalists, corporations have taken great care not to allow their environmentally destructive practices to reach the public eye. Instead, they work carefully to cover their caustic policies with a thin coat of greenwash. Oil companies declare their love of nature and concern over global warming. They tout 'cleaner burning' fuels and decorate their filling stations with green suns. Chemical companies that pump hundreds of tonnes of toxins into the environment affirm their commitment to public health. Products that are slightly less caustic ('cleaner-burning' gasoline) are promoted as green, despite their dire environmental consequences. Truth is an early victim as corporations push their environmental records while they meet only the minimum regulatory standards and fight aggressively against further protections. They highlight the recyclability of their endless excess packaging, or declare their concern for climate change as they block support for alternative fuels and lobby against tighter emissions reductions. Policymakers dutifully do their part, rejecting labeling requirements, opposing tighter health and safety regulations, and allowing every opportunity for 'public–private' partnerships that legitimize corporate claims to environmental responsibility.

This sort of deception and hypocrisy is far from limited to the private sector. The environmental record of George W. Bush, the self-described environmental president, makes this amply clear. The Bush administration's 'Clear Skies Initiative' – a name worthy of Orwell – gave a free hand to polluters. Current air pollution standards, enacted in 1977, included a compromise that grandfathered in coal-fired plants. Environmentalists had hoped that these most polluting facilities would be required to clean up their emissions. However, corporate advocates in Congress argued that such requirements would be too costly. And so, in a provision known as New Source Review, plants were allowed to continue to operate and perform routine maintenance without having to install cleaner technologies. But once they were ready for major upgrades, they would be required to incorporate cleaner equipment. The grace period was expected to last a few years and seemed to be a reasonable compromise. However, for the next three decades, plants continued to get around these requirements by calling even extensive expansions routine maintenance. Upon taking office, the Bush administration moved to do away with this compromise all together. Industries were let entirely off the hook. Even facilities that had already agreed to cut emissions were relieved of the requirement. In a perverse bit of doublespeak, President Bush labeled this plan his 'Clear Skies Initiative' and promised that it would 'dramatically reduce pollu-

6.3 / Serious backsliding: comparing the Bush administration 'Clear Skies Initiative' to existing restrictions

	Nitrogen oxides (NO_X)	Sulfur dioxide (SO_2)	Mercury (Hg)
Clean Air Act	1.25 million tonne cap by 2010	2 million tonne cap by 2012	5 tonnes/year by 2008
'Clear Skies'	2.1 million tonne cap by 2008; 1.7 million tonne cap by 2018	4.5 million tonne cap by 2010; 3 million tonne cap by 2018	26 tonnes/year by 2010; 15 tonnes/year by 2018

Source: Sierra Club, www.sierraclub.org.

tion from power plants.' In fact, Bush policies will allow 450,000 more tonnes of nitrogen oxides, 1 million additional tonnes of sulfur dioxide, and nearly 10 million extra tonnes of mercury into the atmosphere. In addition, although mercury has been shown to cause brain damage, lung and kidney damage, reproductive health problems, and even death, the Bush plan proposed a market-based trading of mercury pollution credits. Since such pollution trading schemes can create hotspots, it's likely that the proposed 190 per cent increase in mercury emissions would be largely imposed on communities of color and the poor.

The examples of manipulation and hypocrisy seem endless. Among his first actions, President Bush moved to increase the allowable level of arsenic in drinking water. The very institutions charged with protecting the environment were stacked with former lobbyists and lawyers from the mining, timber, and chemical industries, all with a clear hostility to environmental regulations. The Bush 'Healthy Forests' program opened up pristine forests to aggressive logging and undercut national conservation efforts. An administration proposal attempted to redefine the meaning of 'Waters of the United States' to exclude up to 60 per cent of the nation's rivers, lakes, and streams from the protection of the Clean Water Act. In carefully calculated terms, President Bush promised to reduce emissions 'intensity' by 18 per cent by 2012 – this translated into roughly a 14 per cent growth in emissions, or essentially business as usual. While the administration repeatedly (but in careful terms) claimed its environmental credentials, it seemed anxious to undercut environmental protections at every turn.

All of these moves and others, from opening up the earth's largest temperate rainforest to logging to abandoning wetlands protections, were carefully orchestrated to remain out of the public eye. Some were

6.4 / A look at some Bush administration appointees

Appointee	Position	Ties with industry
Andrew Card	Chief of Staff	Former president of the American Automobile Manufacturers' Association. Chief lobbyist for General Motors.
Condoleeza Rice	National Security Advisor	Former Chevron board member (Chevron briefly named an oil tanker after her)
James Cannaughton	Chair of the Council on Environmental Quality	Lawyer for General Electric and Atlantic Richfield in their suit against the EPA to avoid cleaning up heavily contaminated sites.
Gail Norton	Secretary of the Interior	Former lobbyist for NL Industries, a major chemical company accused of exposing children to lead poisoning. National chair for industry group representing timber companies and chemical manufacturers
Linda Fischer	Deputy Administrator, EPA	Former vice-president of government affairs at Monsanto, a leading pesticide and GM company.

Source: League of Conservation Voters, www.lcv.org.

announced late on Friday, to avoid press coverage. Some were buried in administrative doublespeak. When the political fallout was potentially too costly, the administration encouraged private groups to do the dirty work. In so-called 'sweetheart suits' the White House encouraged lawsuits against the federal government and then acted quickly to negotiate a settlement in favor of industry. Such calculated deception undercut the very fabric of democratic accountability. Distracted by wars abroad, terrorism, and the administration's flag-waving, a nation of self-described environmentalists seemed to pay little attention to Bush's war on the environment.

The reasons for these policy moves were obvious to anyone who cared to look. The Bush administration's anti-environmental agenda was bought and paid for with tens of millions of dollars in campaign contributions from polluting industries. Mining, oil, gas, timber, chemical, and coal industries contributed more funds to the Bush–Cheney ticket than to all Democratic candidates and party committees combined. A full 77 per cent of the campaign contributions from major polluters went to Republican candidates or party committees in the 2000 election. The investment paid off. After providing $3 million in campaign support, the mining industry enjoyed the elimination of major environmental controls on coal and hardrock mining. After contributing about the same amount, the timber industry enjoyed new policies that encouraged deep logging in national forests. After donating over $18 million to the Bush campaign and party, chemical manufacturers benefited from a policy change that required taxpayers, not polluting industries, to pay for the cleanup of hazardous waste sites. After receiving a donation of $17 million from the oil and gas industry, the Bush administration opened up environmentally fragile areas in and around national monuments and parks to oil and gas drilling.[31] As one 'pioneer' Bush–Cheney contributor (having raised over $100,000) from the coal industry had it, 'We were looking for a friend and we found one in George W. Bush.'[32]

Nevertheless, we were in trouble before George W. Bush took office and, if we do not effect a major course change, we'll be in trouble long after he is gone. Neither Republicans nor Democrats seem to possess the courage needed to challenge the rules of the game. Democratic presidential candidate Bob Kerry railed against Bush policies while simultaneously promising to lower gas prices. What is most disturbing is what is left out of the environmental debates. Policymakers rarely even mention conservation strategies, renewable energy, public transit, or any other fundamental issue that might have a more serious long-term impact on environmental quality. The idea of curbing consumption or placing serious restraints on industry is simply beyond the pale. By its own assessment, the EPA has not focused its efforts on those issues that pose the greatest risk to our shared future. An EPA self-study indicated that the agency has focused on simple, short-range public health rather than the protection of the greater ecosystem.[33] So, for instance, instead of protecting intact habitats when they are healthy, we enforce last-ditch and often hopeless measures to protect vital species only once they're endangered. The narrow focus of our environmental policies and reliance on so-called 'end of pipe' controls (policies that do not address the source or cause of pollution, but only try to limit its release after its produced) virtually ensure failure.

Taking a step back and considering the global picture make the inadequacies of environmental policies all the more clear. Even if we were to empower the EPA to ensure the kind of protection that is necessary, protecting only 'our' water or 'our' air is simply not possible. We may be able temporarily to manage cleaner parks and safer water, but, given the global nature of the environmental crisis, wherever the pollution is, in the end it will affect us all. Addressing the crisis in planetary terms, which in the end is the only way to think about these issues, presents even greater challenges.

Lots of talk, little action

It should be clear by now that the task of saving the planet and ourselves cannot stop at political borders. This realization inspired the first comprehensive international conference on the environment in 1972. Nations from around the globe attended the Stockholm Conference on the Human Environment. The agenda was largely set by wealthy countries and framed by the growing belief that humankind may be reaching the limit of the earth's resources. Widely read books like *Limits to Growth*, published in the same year, had argued this point strongly. Many in impoverished countries worried that this line of reasoning would in the end be used to argue that there was no room left for their development. It seemed to many in the impoverished world that wealthy countries, having now secured their privileged global position, were turning their attention to the protection of nature; and having destroyed their own natural resources in the process of development, the protection of nature seemed to mean placing sharp limits on the capacity of the world's poor to pursue their own development.

Centuries of imperialism and exploitation, and an environmental agenda that was clearly set by the wealthy, gave impoverished countries reason for suspicion. The fascination of relatively wealthy suburbanites with pandas, whales, or other charismatic species inspired environmentalists to pay attention to developments beyond their own borders; yet too often that attention was short-lived and narrow. The projects and proposals of environmentalists in wealthy countries rarely fully recognize the dire straits of the world's impoverished or the inextricable link between global poverty and environmental protection. The world's poor countries started to get nervous. In the decade before the Stockholm conference, the G77 (the group of seventy-seven impoverished countries that refused to align themselves with either superpower during the Cold War) passed a series of UN resolutions asserting their right to develop and the need to leave environmental issues to national governments. A resolution six months before the conference pointed out that global environmental troubles were 'caused primarily by some highly developed countries, as a result of their own high level of improperly planned and inadequately coordinated industrial activities' and that 'most of the environmental problems existing in developing countries are caused by their lack of economic resources.'[34] But, like the impoverished world's demands for economic equity and a New International Economic Order, this resolution fell on deaf ears in the world's wealthy countries.

At Stockholm, any hope that those fighting for the protection of species and forests might find common cause with those fighting poverty and hunger was sorely disappointed.[35] The United Nations Environmental Program was formed. The notion of the Common Heritage of Mankind was institutionalized, and used to declare the oceans, the seabed and space to be common resources of all humanity. Monitoring networks were established to observe global pollution and document the status of natural resources. That was about the extent of it. The moves were primarily about monitoring environmental destruction rather than stopping it. And virtually no measures addressed the link between environmental degradation and economic inequity. While these actions may have been necessary first steps, they were at best baby steps.

Over the course of the following decade, this link was increasingly recognized, and common ground between environmentalists and advocates of international development was becoming more apparent. Part of this common ground was voiced in the idea of 'sustainable development,' a notion first articulated by the World Commission on Environment and Development in 1987. The idea, in an often-repeated phrase, was to meet 'the needs of the present without compromising the ability of future generations to meet their own needs.'[36] This notion recognized two fundamental truths: the kind of development that has taken place in wealthy countries cannot be duplicated in the impoverished world without grave environmental consequences; and we cannot address global environmental problems unless we also take on the entrenched structure of global poverty. This was an encouraging shift in rhetoric, but an associated shift in policy was not as simple.

By the early 1990s, the severity and human costs of the global environmental crisis were growing clearer, and leaders in the impoverished world that had once rejected environmentalism as a distraction from the urgent program of economic development started to sound more green. The forest resources on which more than 200 million people were dependent were increasingly at risk. Land degradation threatened the livelihood of perhaps 800 million. Lack of adequate food, water, and sanitation affected about a third of the earth's people. Air pollution was growing more and more severe in cities around the impoverished world, leading to the deaths of 35,000 children *every day*.[37]

It was in 1992, at the United Nations Conference on Environment and Development in Rio de Janeiro – the 'Earth Summit' – that the link between international poverty and environmental degradation started to gain more attention. The culmination of two years of preparatory meetings, the conference greeted representatives from 178 countries and more than a hundred heads of state. Under the broad theme of sustainable development, delegates discussed and argued the loss of biodiversity, climate change, deforestation, and soil degradation, among other topics. A clear enthusiasm was in the air. A celebratory spirit prevailed among activists and delegates alike. Speeches, concerts, and dancing around a bronze statue of the Tree of Life filled the meeting with a spirit of solidarity and hope. But, once again, when all was said and done, there were no significant policy changes, no remarkable shifts in international law, no adequate global programs to defend the earth. Still, many felt that a 'paradigm shift' had been marked. If we hadn't defined a solution, at least we seemed to have a better understanding of the problem, they argued. Maybe.

What is undoubtedly true is that the agenda of the meeting had been clearly shaped by the world's wealthy. The US and other powerful nations were inclined to cast the idea of sustainable development in the narrowest terms possible. Discussing changes in the global economic system, a prerequisite to real sustainability, was simply not going to happen; and even the link between environmental sustainability and economic justice would not be directly addressed. In fact, the US spent a fair bit of time opposing the use of 'Environment *and* Development' in the title of the conference, proposing the title 'Earth Charter' instead, since this suggested a more narrow focus on environmental protection rather than economic and environmental equity.[38] The issues of most concern for impoverished countries – how to provide safe drinking water or stem land degradation – weren't even addressed at Rio since they were of little interest to the leaders of wealthy countries. Hence, despite the enthusiasm of many participants, that same fundamental conflict of interest between rich and poor continued to block real progress.

We will discuss the negotiations on climate change in greater detail shortly, but a brief mention here helps illustrate this conflict. The world's poor countries argued that the wealthy, industrialized countries needed to clean up their act as the first step in combating climate change. Impoverished countries do not produce the bulk of greenhouse gases, they argued, and, particularly given their struggle to develop, they should not be asked to sacrifice the scarce opportunities they have for development to deal with a problem that is primarily caused by the wealthy. It should be no surprise that wealthy countries, notably the US, disagreed. US representatives argued that although the US produced a full quarter of the world's greenhouse gases, the rapid expansion of carbon dioxide emissions and of population in many impoverished countries required that they also commit to firm reductions.

Wealthy countries also wanted to tie discussion of climate change and emissions to discussion of the preservation of forests. Forests, they argued, functioned as sinks for greenhouse gases and must be an integral part of the climate change calculus. Impoverished countries were wary of being pressured to preserve forests, and thus limit their use of resources and their potential development, in order to counterbalance the excessive emissions of the wealthy. India and Malaysia argued that attempts by international organizations to manage their forests would be an infringement on their sovereignty, and wanted discussions on emissions and forest protection separated. If they had to be linked, they argued, then since preserving forests was a benefit to the global community, the costs should be shared by other countries and not be the sole burden of the forested country. They called for compensation to countries with large forests to help offset the cost of conservation and, more importantly, to offset the lost income from not destroying these forests.

Debates over biodiversity and biotechnology also revealed the cleavage between rich and poor. Impoverished counties with rich and diverse tropical forests saw these as a national wealth that needed to be protected from transnational corporations. They knew what it meant to have corporations 'mining' their forests for biological material and privatizing what had previously been treated as a shared resource. They pushed for a biodiversity treaty that would ensure their stake in their natural resources (and thus, they argued, their interest in the preservation of biodiversity) and proposed that access to genetic resources be subject to the 'prior informed consent' of the countries where the resources are found. The benefits, they argued, should then be equitably shared with that country on 'mutually agreed terms.' This suggestion received broad support.

Unfortunately, this proposal threatened the profit margins of Western corporations, many based in the US. Indeed, the US was the only country at the conference that did not sign the resulting Convention on Biodiversity. Within five years, 169 countries would ratify the treaty. President Clinton would eventually sign the agreement in 1993, but only after being given the green light by key industries, and only after negotiating a serious weakening of the treaty's restrictions. With the urging of the Clinton administration, key corporations such as Merck, Genentech, and ADM had come to see this compromised treaty as facilitating potentially lucrative cooperation with governments around the world. But not everyone was pleased. Thirty-five Republican senators stood firm to block the treaty, voicing their concern for its impact on US business and sovereignty.[39] This made the required two-thirds support for ratification impossible. As a result, the US was relegated to observer status in further negotiations, weakening both the importance of the agreement and the US capacity to shape its final form. Partisan politics and interest brokering had trumped even the weakest attempts to deal with the increasingly dramatic loss in the world's biodiversity.

The eventual agreement on forest protection that emerged from Rio was similarly hobbled, and designed to suit the interests of the global timber industry. The resulting document, the awkwardly phrased 'Non Legally Binding Authoritative Statement of Principles,' while filled with green promises, continued to push free-market access for timber interests.[40] Article 13 called for 'open and free international trade in forest products.... [And the] reduction or removal of tariff barriers and impediments to the provision of better market access and better prices for higher value-added forest products and their local processing.' The loss of forests around the world continued; and the laudable declarations against deforestation made at the Earth Summit and after proved to be little more than words.

The conflict between rich and poor permeated all areas of discussion. Wealthy countries, increasingly likely to see the impoverished world as a place to dump their toxic trash, argued that market forces should determine where waste is disposed of. Impoverished countries argued that wealthy countries should begin by reducing the tremendous amount of hazardous waste they produce and not seek to dump it on the impoverished. Population issues defined another arena of conflict. Impoverished countries did not want to discuss the sensitive issue of population growth. Leaders and some environmentalists in wealthy counties were more likely to argue that global population growth was a major environmental threat. In particular, some argued, population growth in some larger countries like China, where consumption and auto use were growing, could have important consequences for global climate change. Leaders from the impoverished world often responded that the overconsumption of the wealthy countries was the real issue. And discussion of population growth in poor countries, where the per capita impact on the environment was relatively slight, was a distraction. And, they added, the desire of wealthy countries to focus on population reflected a general effort to place the blame and burden of environmental protection once again on the backs of the world's poor.

Any effort that tried to expose the link between environmental threats and the global economic system were shot down. There was no serious discussion of international debt or poverty and trade. China and India declared their refusal to limit their efforts to develop in order to deal with a climate issue that they saw as caused by the US and Europe. The US, and to a lesser extent European countries, precluded any proposal that sought to limit continued economic expansion of the wealthy or curtail their access to markets and resources; and, more generally, they refused to recognize any conflict between continued growth and environmental sustainability.[41] With similar effect, oil-producing nations, powerful industries, and the representatives of wealthy countries teamed up to reject any move that might limit the continued use of fossil fuels.

US delegates insisted on discussing only what they termed 'win–win' solutions. That is, they rejected the idea that environmental protections required a slowdown in consumption and profits, and were only willing to discuss policy changes that would not hurt the corporate bottom line. This, of course, meant no real changes would be discussed. Because, while it is certainly true that cleaner production and waste minimization can produce win–win outcomes (less environmental damage without threatening profits), this ignores the bigger question. Saving the earth from global warming, toxic overload, or deforestation is not going to be accomplished by gradual improvements in efficiency and better waste management. The fundamental stucture of the global economy is at the heart of the problem. A narrow focus on so-called win–win solutions guarentees an eventual lose–lose–lose outcome – the poor, the rich, and the earth will lose in the long run.

So the results of the Earth Summit were largely symbolic. Agenda 21, a several-hundred-page program for sustainable development, was adopted with quite a bit of fanfare. Beginning with the declaration that 'Humanity stands at a defining moment in history,' the proposal addressed social and economic development, resource conservation and management, as well as the empowerment of women, indigenous groups and others. It recognized that sustainability cannot be achieved in a world with extreme poverty, vast inequality, and local communities that are excluded from access to or management of their natural resources. It declared that world poverty is the shared responsibility of the entire global community. It outlined strategies for protection of the atmosphere, combating soil degradation and deforestation, the use of toxic chemicals, and the disposal of hazardous waste. The moral responsibility of the world's wealthy to assist the world's impoverished was recognized. Wealthy countries were to commit to a sustainable level of consumption, reform their economies, and internalize environmental cost in the price of goods. A host of admirable tasks and global reforms were to be headed by a proposed Commission on Sustainable Development. All are very good ideas.

Unfortunately, in the end these were mostly just words. No binding commitments were included, and the goodwill generated by the conference quickly dissipated. Although the 128 representatives of the G77 at Rio pushed for the creation of stronger moves toward sustainable development, the US and other wealthy countries refused to commit significant resources to support Agenda 21 priorities. None of the resulting agreements defined binding international requirements, much less a substantial change in business as usual. Rather than address the fundamental questions of economic growth and ecological sustainability, leaders debated minor variations in phrasing and issued impressive-sounding but toothless resolutions. They settled for the politically palatable rather than achieving the environmentally necessary.

Sometimes empty symbolic measures can actually have negative consequences. The United Nations' effort to encourage environmental responsibility among corporations is an example. Corporate signatories to this 'Global Compact' were expected to commit themselves publicly to nine principles associated with environmental protection, labor standards, and human rights; but the UN resisted any call for independent monitoring. Of course, with no monitoring of corporate policy by the UN, endorsing the Global Compact became an easy way for industry leaders to proclaim their environmental righteousness – with apparent endorsement from the UN – without actually changing much at all.[42] Corporate leaders jumped at the opportunity to co-opt the language of the Earth Summit. In the end, the initiative did much more to confound public concern and obscure environmental responsibility than to encourage any real change in corporate practices. Advocates argue that voluntary initiatives and corporate self-regulation could be more effective than actually limiting corporate behavior. The trouble is, without independent enforcement or monitoring, such voluntary program can undercut public pressure toward responsible environmental, human rights, and labor practices while offering nothing in return. Troubled industries got a bright green polish on their public image, courtesy of the UN, without altering their destructive practices.

Subsequent follow-up conferences, the so called Rio + 5, five-year follow-up negotiations, and the Rio + 10 conference in Johannesburg reflected the same impotence that was evident at the 1992 conference. World leaders in Johannesburg patted themselves on the back and many made impassioned appeals for greater attention to issues of global poverty and sustainability; most agreed that the Earth Summit process had stimulated ongoing debate, but no significant new actions were taken. The US, for its part, hardly acknowledged the event. President Bush snubbed the Johannesburg conference; and his representatives teamed up with delegates from oil-producing states and the fossil fuel lobby to squash efforts to set serious targets for the development of sustainable energy sources. The US even sought to have all mention of global warming removed from the conference's plan of action! Even the most sympathetic member of the US delegation, Secretary of State Colin Powell, received boos and heckles when addressing the delegates. So, in the end, thousands of delegates from around the world converged on a spruced-up conference center in a wealthy neighborhood of South Africa. They toured a plush shopping mall and were shuttled about in luxury automobiles, all to reconfirm their environmental righteousness for the world's television cameras. The final declaration of principles revealed some slight progress on a few issues but also some backsliding on others. The idea of serious binding commitment was rejected and previous promises of assistance to impoverished countries were left unfulfilled.

Supporters will tell you that the Earth Summit, Agenda 21, and subsequent declarations and meetings reflect an important cooperative effort and voiced emerging global hopes and expectations for the first time. In this way, they set the agenda for future discussion and thrust sustainable development on the world stage. Many claimed that activists, leaders, and even business interests defined new partnerships and recognized new alliances at Johannesburg. Environmental organizations began to see the full importance of addressing poverty in their drive for environmental protections; and advocates and leaders for impoverished countries more clearly acknowledged the centrality of environmental issues to their aims. Perhaps the groundwork for future, practical action was laid there. Perhaps it was not. Meanwhile, the war against nature continued.

Ozone: a small step in the right direction?

If moves toward a general, overarching agreement on the global environment have not yet been entirely successful, there have been some measured successes on more focused problems. A popular example is the international negotiations to stop the destruction of the ozone layer. Many point to the resulting agreement as a model for other environmental issues such as global warming. Yet the success of this agreement is in a small category – a category of one to be precise. The uniqueness of the causes of ozone deterioration and the particular shifts in industry policy that allowed agreement make this an unlikely model for action on climate change. So, as we'll see, while this success case may give us a bit of room for hope, it also reminds us of the powerful forces working against global environmental reform.

A major loss of the ozone layer over Antarctica was first noticed in 1982. By the end of the decade, ozone levels in some Antarctic locations were down 60 per cent from their readings two decades previously. Some areas had almost no ozone at all; and each spring this hole in the ozone layer grew bigger. By the beginning of the 1990s it was about the size of North America.

Scientists figured out fairly quickly that it was our fault. Human-made halons from fire extinguishers and CFCs from aerosol cans were to blame. As we saw in the first chapter, the highly reactive chlorine and bromine that these chemicals release into the atmosphere destroy ozone at high altitudes. Although these chemicals had been in use since the 1930s, the use of CFCs had increased significantly. By the late 1980s, well over 1.2 million tonnes were being produced annually. Along with aerosol spray cans, the product was used in air conditioners and the manufacture of Styrofoam. As a result, the amount of chlorine in the atmosphere had reached four and a half times its natural level.[43] We were losing about 5 per cent of the ozone layer each decade – not a huge loss perhaps, but clearly not sustainable. Within a few decades we would be in big trouble. Once again, the source of the problem reflected global inequalities: a little over a quarter of these dangerous chemicals was produced in the US and only about 4 per cent was produced in impoverished countries.[44]

With the announcement that the ozone decline was caused by CFCs, the public reaction was notable. Purchases of aerosol sprays declined significantly, prompting some companies to phase out their aerosol products. Under popular pressure, lawmakers in the US moved to restrict the use of 'nonessential' CFCs in 1978. Within a short period, CFCs in aerosols were reduced by 95 per cent in the US. Some European countries followed suit. After this initial reaction, however, world CFC use once again began to rise. New applications of CFCs in foams and solvents reversed the decline initiated by the aerosol ban. So, despite the clearly recognized environmental damage these substances posed, industry turned a blind eye. CFC production surpassed pre-ban levels by 1984. Use of other ozone-depleting chemicals also continued to increase.

In spite of these warning signs, the Reagan administration helped ensure the continued growth of the CFC industry. The US rejected a group formed by some European countries and Canada that called for an international ban on CFCs. Instead US industries formed the perversely named 'Alliance for Responsible CFC Policy,' which waged a powerful and well-funded campaign against regulation; and the pro-industry Reagan administration proved happy to oblige this new group. In a disturbing preview of later claims by industries regarding global warming, manufacturers of CFCs argued that the scientific evidence was not strong enough to justify drastic measures and that the economic cost of reducing CFCs would be too high. The Reagan administration fell in line, rejecting all but the most modest efforts to reduce CFC use. Since the US was the world's major

6.5 / Emissions of ozone-depleting chemicals in the US (million tonnes of ozone-depleting potential)

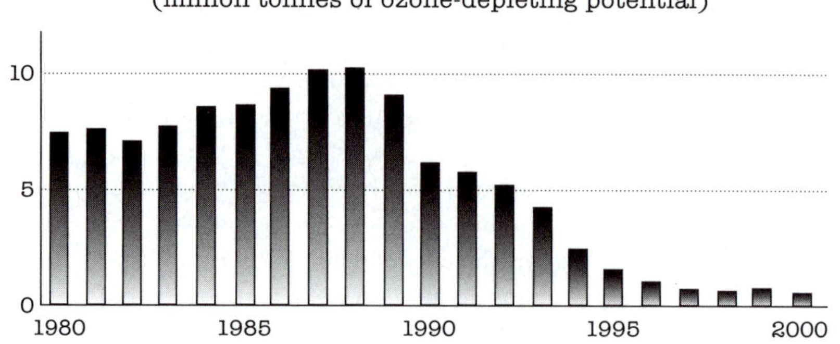

Source: Alternative Fluorocarbons Environmental Acceptability Study (AFEAS), www.afeas.org.

manufacturer and user of ozone-depleting chemicals, if it did not take the lead any efforts were likely to have little global effect.

However, in a short time, US policymakers sharply changed their position on ozone-depleting chemicals. The EPA advocated a full phase-out of CFCs in 1986. What changed? Simply put, corporate interest changed and US policy followed suit. DuPont, the major US producer of CFCs, controlling 50 per cent of the US market and a quarter of the global market, had begun researching substitutes for CFCs in the 1970s when sales began to slump. Because the company moved on alternatives to CFCs before its competitors, any ban on their use would give the company a sharp advantage. With such a large share of the market, DuPont's shift meant an industry-wide shift. With the green light from the industry leader came a change in the position of the administration and key lawmakers. A resolution passed by the Senate with only two voting in opposition called for a 95 per cent reduction in CFCs within a little more than a decade.

With industry opposition ended and public concern heightened by the discovery of the Antarctic ozone hole two years earlier, a landmark international agreement was forged. Britain and France, the homes of Europe's largest CFC producers, who had not developed the same lead on alternatives as DuPont, were still strong holdouts. However, advocacy for CFC reductions by other European countries led to a European agreement to move toward a 50 per cent reduction in CFCs by the end of the century. In 1987, with the US now taking a leading position, the Montreal Protocol codified this aim. Developing countries, which consumed very low amounts of CFCs, were given ten extra years to comply with the agreement. Technological assistance in developing alternatives to CFCs

6.6 / Concentration of ozone-depleting gases in the atmosphere holding mostly steady

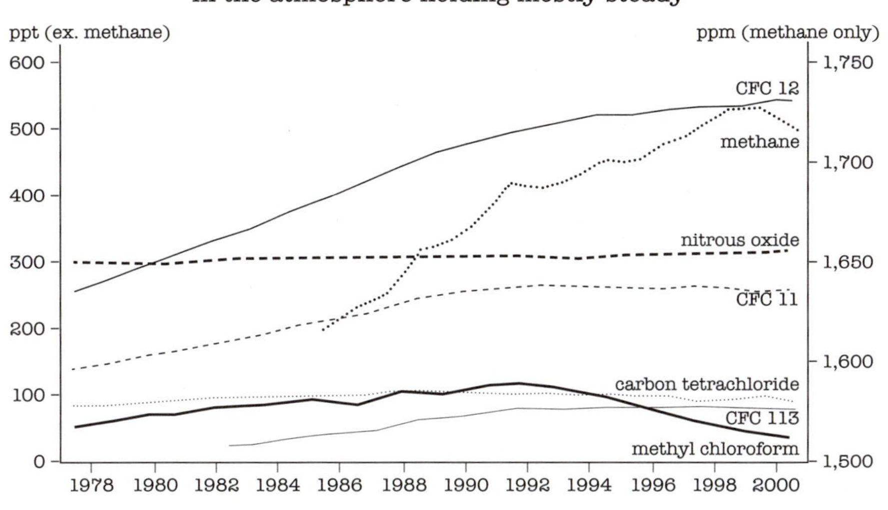

Source: Source: *Earth Trends*, World Resource Institute, earthtrends.wri.org.

was also promised. A year later, when the Ozone Trends Panel reported that the ozone layer was receding faster than previously thought, the inadequacies of the Montreal agreement were clear. Public pressure on governments and CFC producers increased. With alternative products ready to move, DuPont announced that it would stop producing halons and CFCs by 1999. Under similar pressure, British producers followed. The initial Protocol was followed in 1990, 1992, and 1995 with increasingly immediate reduction targets, leading to a dramatic drop in CFC and halon production by 1999.

However, even with this largely successful treaty, the problem of ozone depletion was not entirely resolved. While sharply reduced, CFCs are still being produced. In 2001, nearly 30,000 tonnes of CFCs were produced globally.[45] The key chemical that replaced CFCs, HCFCs, produces only about 5 per cent of the damage to the ozone layer as CFCs. However, because we produce a lot of it (500,000 tonnes of HCFC were produced in 2001), the ozone-depleting potential is still there.[46] An attempt to accelerate the phase-out of these chemicals has been firmly rejected by industry groups.[47]

Even more disturbing, chemicals used in solvents and adhesives, chemical feedstock, pesticides, and agricultural fumigants that have been identified as ozone-depleting have not been banned. Without equally profitable alternative chemicals, industries have fought hard to defend the continued use of these chemicals.Arguing that a ban would be too expensive, representatives of US grain industries have successfully rolled back movement on the elimination of one such chemical, methyl bromide. US grain mills use about 60 million pounds of methyl bromide every year, mostly to fumigate grain mills, silos, and soils.[48] The US accounts for about 40 per cent of global use, but its share is declining. A ban on the pesticide would increase the time it takes to fumigate a silo from two to five days – and time is money. On the other hand, the UNEP estimates that a ban on methyl bromide would reduce damage to the ozone layer by 4 per cent almost instantly. This would be a 'major benefit' during the next decade, when the recovery of the ozone layer is at a critical stage.[49] The Bush administration predictably sided with industry against the environment, pushing for a 'critical use exemption' so that US industries can continue to use the ozone-depleting chemical.[50]

What lessons can we learn from the battle to save the ozone layer? First, it's clear that corporate interests had a powerful influence on setting the agenda and defining policy throughout the process. When industry leaders did not want action, no action occurred. When industry leaders supported action, action became possible. And public outrage and a shift in buying patterns can have an important impact on corporate and public policy when clearly focused and defined; but corporations will resist any change that affects their bottom line. The role of the US, as the world's most powerful nation, was crucial. If the US takes a lead, even with international resistance, meaningful reform is possible. Without US leadership, real reform is unlikely. And US leadership, as always, hinges on the demands of industry.

Basel Convention: a model of successful resistance?

There is room for optimism. Make no mistake, the entrenched influence of corporate power brokers is formidable. But a coordinated and determined grassroots resistance can force change. The potential for global solidarity toward change was made clear by the international effort to end the export of toxic waste to impoverished countries.

Through the 1970s and 1980s, disposing of toxic waste became tougher and tougher in the US and Europe. In upstate New York, the Love Canal neighborhood, built over a former chemical waste dump, was the first man-made disaster to be declared a national emergency. As people grew more sharply aware of the risk of hazardous waste dumping, they

6.7 / Known obsolete pesticide stockpiles, selected African countries (tonnes)

Ethiopia	1,924
Sudan	1,097
D.R. Congo (Zaire)	591
Mali	238
Ivory Coast	825
Cameroon	225

Source: Greenpeace, *Pops in Africa*, 2000.

demanded regulations to reduce and control what had been irresponsible disposal with often tragic consequences. Toxic waste dumps were closed. Waste disposal was more responsibly regulated. And as a result the cost of waste disposal grew.

With hundreds of millions of tonnes of hazardous waste produced in the world's wealthy countries every year, and options for toxic disposal growing more expensive and unpopular, industries looked to the impoverished world as a place to dump their poisonous by-products. Waste that could cost $3,000 per tonne to dispose of in the US could be dumped in Africa for a pay-off of as little as $2.50 per tonne.[51] Countries with the strongest environmental regulations became the largest exporters. Industries in Canada, Germany, the Netherlands, and the US regularly shipped tonnes of toxic waste to the poorest states in Africa, Central America, and Asia. Waste followed the path of lowest costs. The world's most vulnerable communities, crippled under the weight of mounting poverty, once again felt the burden of injustice as the dangerous waste of the world's wealthy was dumped in their midst. Corrupt regimes in destitute counties were all too ready to gain short-term pay-offs by facilitating the dumping.

By the early 1980s, headlines began appearing regularly detailing the discovery of hazardous waste and chemicals that had been dumped on beaches, the countryside, and villages around the world. Infamous 'ships of death' like the *Khian Sea* gained international news coverage. The *Khian Sea* left Philadelphia with a permit to deliver fertilizer to Haiti. After nearly a quarter of the 13,000 tonnes were unloaded, it was discovered that the actual cargo included waste containing highly toxic chemicals like dioxins and furan and heavy metals such as mercury, lead, cadmium, and arsenic. When ordered to reload, the ship disappeared in the night, leaving 4,000 tonnes of highly toxic ash on the Haitian beach, where it would sit for nearly a decade. The ship spent two years trying to find a dumping ground for the remainder of its deadly cargo. Finally, it set off through the Mediterranean and around the coast of Africa; when it arrived in Singapore, it had a new name and an empty hull, having presumably dumped its toxic cargo in the open sea.

Impoverished countries had neither the resources not the expertise to deal adequately with such waste. Even in wealthy countries, landfills were often not safe and dangerous waste often polluted rivers and lakes and poisoned local populations.[52] In the impoverished world, disposal sites were far less regulated, if they were even clearly identified. Waste was often dumped in leaking containers, or in no containers at all. Toxic chemicals leached into the soil and water. In Koko, Nigeria, for example, nearly 4,000 tonnes of highly toxic PCBs and asbestos from Italy were left in leaking drums by a local industry that had forged documents and paid bribes to arrange the dump. Hundreds of mysterious barrels washed up on the Turkish shore. When curious locals opened the barrels, they suffered nausea and rashes. These stories have been repeated again and again throughout Africa, Asia, and Eastern Europe.

Sometimes the waste has been disguised. It could be labeled as fuel, for example. Or contaminated soil could be shipped as fill dirt for roads. At times, toxic waste was even labeled as foreign aid! In a particularly infamous case, three South Carolina companies conspired to dump waste by mixing 1,000 tonnes of ash containing high levels of lead and cadmium with 3,000 tonnes of fertilizer for export. Before the scheme was uncovered, farmers in Bangladesh had spread it on their fields, exposing themselves, their families, and the local community to dangerous poison.

There are many obvious reasons why such practices are reprehensible. Clearly, the death of unsuspecting farmers and others who are poisoned by our waste is hard to justify in any terms. The fact that the poor countries do not have adequate resources to deal with this waste in an environmentally sound manner adds to this argument. Moreover, the international export of waste undercuts the potential of policies like the Clean Air Act and Clean Water Act. These laws are supposed to do more than protect our backyards alone. Raising the cost and barriers to the disposal of waste creates a powerful incentive for industries to produce less waste, or produce waste that is less dangerous to humans and our ecosystem. That is a key value of such laws. But exporting waste can be an easy escape for polluting industries, allowing them to undercut the intent of the law.

In the late 1980s, African states led an effort to ban this 'toxic colonialism,' catalyzing an international coalition that included several European parliamentarians. The US led the opposition to this effort, arguing that any restrictions on the export of toxic waste would interfere with free trade. Instead, US negotiators proposed an informed consent policy, which would end only the very worst kinds of dumping – secret shipments dumped on the countryside without the knowledge of the local population or their government. African states demanded more than this, and proposed amendments to require inspections of disposal sites by the UN and prevent exports to countries that lacked adequate facilities to deal with the waste.

African states knew that it was most often desperation not ignorance which led impoverished countries to accept toxic waste. When the government of Guinea-Bissau briefly agreed to accept 15 million tonnes of toxic waste for $600 million (four times the total value of all goods and services produced in the country for a year) the reason was not because they did not know the waste was dangerous and likely to lead to severe illnesses and premature mortality. It was because they were worried about survival *today* and felt that they could not afford to worry about tomorrow. For many impoverished countries the choice seemed to be die now from starvation or die in a few years from cancer and other illnesses brought on by the toxins buried in the soil by multinational corporations. From this perspective, African countries argued that only strong and encompassing regulations could effectively end this lethal endgame. Industrialized countries, concerned that their access to cheap dumping sites would be lost, waved the sacred flag of free trade and rejected this proposal outright.[53] From this perspective, the world's impoverished had a new commodity, something they could sell on the global market to pay their rising debt: they could sell off their health and the health of their ecosystem.

Initially, the world's rich and powerful and the perverse logic of free-market economics had their way. The resulting agreement in 1989, the Basel Convention, did not ban the movement of toxic waste but only required that countries know the content of the waste and give written consent prior to dumping. The convention barely went any further than existing regulations that had already failed to control the problem. There was no provision to stop waste shipments that did not meet set requirements. Radioactive waste was excluded. What constituted 'hazardous waste' was not clearly defined. And any formal agreement between two countries was exempt from the restriction. Nevertheless, even this severely weakened agreement was apparently too much for the US, as it rejected the agreement. A frustrated group of African countries walked out of the negotiations, refusing to sign the watered-down, essentially pointless treaty. Within a short time, it was clear that the Basel agreement had very little effect. The global dumping of toxic waste continued to increase.

African governments and environmentalists continued to pressure toxic waste exporters. The environmental group Greenpeace took a leading role in the struggle, publishing a report that listed 1,000 cases of illegal toxic waste dumping. Facing growing popular pressure, some European countries agreed to dispose of their waste at home after negotiations with a determined group of African, Caribbean, and Pacific nations. Several African states agreed to ban all imports of hazardous and nuclear waste to their countries. Six Central American nations did likewise. Pacific nations began movement on similar pacts. Under increasing pressure from Greenpeace's campaign and the demands of African governments, by early 1994 over a hundred countries had agreed to ban the import of hazardous waste.

Even US opposition to a ban seemed to be weakening when, just before a meeting of the Basel Convention's parties in 1994, the Clinton administration announced its intent to ban hazardous waste exports to impoverished countries. However, the proposal carefully excluded scrap metal, textiles, and paper that were traded for recycling. These exceptions are more dangerous than they appear. As Greenpeace representatives forcefully argued at the time, toxic waste could simply be relabeled as recyclables and dumped in sham or inadequate recycling operations. Virtually any sort of waste could be declared to have some sort of 'further use' and therefore be called recyclable. Indeed, the amount of shipments designated as recyclables was sharply increasing. Even in the instances where the waste was in fact able to be recycled, it was still often poisonous. The amount of hazardous waste exported from major industrial countries like the US and Germany increased every year, some of it relabeled as recyclable, some of it redirected to South American or Caribbean countries that had not enacted a ban. After the collapse of the Berlin Wall, Eastern European counties became favorite dumping spots for waste shipments.

At the Basel Convention meeting that followed the Clinton administration's announcement, a remarkable shift was evident: the moral and political outrage prompted by the disposal of toxic waste on the world's poor catalyzed an unprecedented alliance. A broad coalition of countries led by the G77 demanded a complete ban on hazardous waste shipments from wealthy countries to the world's poor. Global toxic dumps had become potent political symbols of exploitation, inspiring determination on the part of several African leaders and the solidarity of China and Central and Eastern Europeans. The resulting coalition of 120 countries pushed forcefully against the economic influence of rich and powerful governments. Despite intense lobbying by waste-exporting countries and promises of future pay-offs, the coalition stood firm.

A number of watered-down counter-proposals were presented by wealthy countries and firmly rejected by the G77 coalition. As the chair of the G77 declared, 'These proposals have loopholes that would quickly widen. We would have a flood of [hazardous waste] movement ... from countries that can cope to countries that cannot.'[54] Not a single impoverished country was persuaded that trade in hazardous waste was a good idea; and no country was successfully bought off with US influence or promises. European countries suggested that impoverished countries that were willing to accept waste could be identified and exempted from the ban. This notion of designated global dumping grounds held little sway among the coalition of countries that were victims or potential victims of toxic dumping. They insisted on no exceptions to the global ban. In the face of this strong determination, European resolve weakened and led to an endorsement of a total ban by key European players. Since the US had not ratified the original Basel Convention, its capacity to influence the proceedings was moderated. So, while US representatives worked aggressively to support European opponents to a ban, they could not successfully squash this effort. The apparent discontinuity with Clinton's earlier public support for an end to toxic exports cost the US a lot of credibility and seemed to many to underscore the lack of dependability in the US position.[55]

Even after the ban was agreed upon, resistance was sharp. The US and Germany rejected the ban, arguing that it had not been strictly adopted as an amendment to the original agreement and therefore was not binding. This objection forced a repeat of the process a year later to enact the agreement as a formal amendment. This time advocates for a ban faced the organized and determined opposition of the US, Australia, and Canada and a powerful and organized corporate lobby. Industry organizations like the International Council on Metals and the Environment fought hard against this ban, claiming that is would be 'GATT-inconsistent and trade-disruptive.'[56] With the potentially lucrative trade in waste threatened, the US dug in its heels. Nevertheless, the threat of massive toxic shipments headed to their shores inspired solidarity among the world's impoverished countries, and the ban was agreed to as a formal amendment to the original Convention.

However, despite this victory, the fight against the global dumping of hazardous waste on the world's poor is still under way. The ban will not officially take effect until it has been ratified by three-quarters of the parties that signed the treaty. This is unlikely to happen soon. At present there is a general agreement to honor the ban, but there is no force of law behind it. If the general momentum of ratification doesn't

6.8 / Selected international environmental agreements

Year	Agreement	Aim	Budget ($ m)	Staff	US position
1973	Convention on International Trade in Endangered Species (CITES)	Restricts trade in threatened or endangered species.	50	27	Ratified in 1974.*
1987	Montreal Protocol on Substances that Deplete the Ozone Layer	With amendments, encourages the phase-out of ozone-depleting chemicals.	3	16	The US played a leading role in defining the treaty, once given a green light by CFC manufacturers. Now working to extend deadlines.
1989	Basel Convention on the Control of Transboundary Movements of Hazardous Wastes	Restricts the export of hazardous wastes from wealthy to impoverished countries.	1.7	17	The US is not a party to the treaty, but has expressed a desire to continue participation in negotiations. The US worked to resist and then weaken the treaty in the past.
1992	Convention on Biodiversity (CBD)	Encourages conservation and equitable sharing of biological resources.	9	50	Clinton signed the convention in 1993, after pressuring for pro-business reforms; but it was never ratified by a Republican Congress.* The US is the major proponent of WTO intellectual property amendment which significantly undercuts this treaty.
1997	Kyoto Protocol to the 1992 UN Framework on Climate Change	Requires reductions in carbon dioxide emissions.	12	100	Signed by Clinton in 1998, after achieving several concessions that reduced the US commitment; nevertheless rejected by Bush upon taking office and very unlikely to be considered by Congress.

* For the US to be a party to a treaty, it must be signed by the president and then ratified by two-thirds of the Senate.

Source: Modified from Hilary French and Lisa Mastny, 'Controlling International Environmental Crime,' in Lester Brown et al., *State of the World 2001*, Worldwatch Institute, 2001, pp. 169–70.

continue, the ban could simply fade away. Australia, the UK, Germany, and Canada are among the countries that have agreed to the ban but continue to work to undermine its integrity. The US is the only wealthy, industrialized country that has not agreed to the ban or ratified the original Basel accord; nonetheless it is aggressively working to weaken it. Industry associations have dedicated hundreds of thousands of dollars to undercutting the ban through legal challenges.[57] Industries have challenged the definition of 'waste,' for example, arguing that hazardous waste with even the most tenuous possible further use should be designated as 'secondary raw materials' not waste, and thus not subject to the prohibition. Because the United States uses a more narrow definition of 'hazardous' than the rest of the world, it is not known how much of the more than 250 million tonnes of waste produced in the US each year is shipped to the impoverished world.[58]

Local communities have paid a high price; examples are too plentiful. Mercury recycling plants in Africa poison their workers and dump high levels of toxins into rivers used by local communities for washing and cooking. Impoverished workers who melt down car batteries in Asia develop lead poisoning. Workers are exposed to deadly fumes when plastics are burned in unregulated and unprotected facilities. Desperately poor boys, girls, men, and women in China scavenge through hazardous electronic waste from the US with no protective gear. Marked as recyclable, the seemingly endless shipments of discarded computers, cellular phones, and other electronic refuse contain lead, mercury, cadmium, and other toxic substances. When the sorters are done, what's left will be burned, buried, or dumped into local rivers, endangering the lives of people in villages, towns, and cities far beyond the local area.[59]

There are also efforts under way to challenge which countries should be designated as wealthy exporting states. Since the ban relies on making a distinction between wealthy sending states and poor receiving states, challenging this distinction could make the entire ban meaningless. Most threatening of all is the looming specter of a challenge to the ban in the WTO. If the WTO authorities find that the ban restricts their sacrosanct free trade, which has been the core of the US opposition, the worldwide movement of toxic waste could once again expand.

Yet there is reason to celebrate. It is clear that the international dumping of toxic waste would be even greater if the ban had not been forged. The victory, even if limited, reminds us that the concerted and determined efforts by exploited peoples, facilitated by global organizations like Greenpeace, can have a powerful effect, and the results can be beneficial for the entire earth.

Kyoto

Perhaps the most urgent crisis threatening our earth and all its peoples is global warming. As the first chapter makes clear, if this is allowed to continue unabated, the results will be nothing short of cataclysmic. The impact of climate change will take its greatest early toll on the world's impoverished; but, eventually, we will all pay a high price.

The trouble is that the greenhouse effect is more difficult to address than the depletion of the ozone layer or the trade in toxic waste, for a number of reasons. First, the production of greenhouse gases is involved in a broad range of industries. It's not just about a relatively few companies, as was the case with ozone depletion. And it doesn't offer a clear villain and a clear victim, as trade in toxic waste did. Wealthy countries are far and away the largest producers of greenhouse gases, but the greatest increase in emissions is happening in some impoverished countries. Second, the impact of global warming is more easily denied or ignored by politicians and corporations because it is more multifaceted than ozone depletion and less acute than toxic waste poisoning. It's too easy for industries, politicians, and uninformed citizens to ignore global climate change until it's too late. Lastly, we are all deeply implicated in climate change. The production of greenhouse gases is tied to innumerable everyday activities, from driving your car to flipping a light switch. Unlike CFCs, there is no easy replacement for the practices that lead to global warming. Dealing with climate change will require us to make a significant change in how we live, not just the kind of paint we use or the chemical in our fire extinguishers.

The problem is not exactly new. Back in 1979, the first World Climate Conference recognized that climate change was a serious threat and launched a series of meetings on reductions. In 1981 President Carter acknowledged the need to address the problem. And after a severe drought in the late 1980s, Congress passed the Global Climate Protection Act, requiring the president to prepare a plan for stabilizing greenhouse gases. However, neither President Reagan nor President Bush was interested in pursuing an issue that seemed so opposed to their aims of continued corporate expansion and wealth. In 1989, environmental ministers from sixty-seven countries met in the Netherlands and agreed to take action to reduce their emissions and assist impoverished countries to reduce theirs as well. But it's clear that any real effort to address global warming would have to be international in scope and would have to include the US. This is so, first, because the US is simply the world's largest producer of greenhouse gases; and, second, because other countries are less inclined to make sacrifices in their own emissions if the US is not willing to commit to some reduction.

The first major international negotiations on the reduction of greenhouse gases took place in parallel with the Earth Summit in the early 1990s. Encouraged by the success of the Montreal Protocol on the ozone layer, the aim was to create a similar 'framework' agreement within which subsequent negotiations could take place. With an air of optimism, many wealthy countries came out in favor of clear and ambitious targets, and they found strong support from the scientific and NGO community. The US, for its part, remained completely opposed to any quantitative targets. Industry leaders and some oil-producing countries rallied alongside the US to head off any commitments to reductions.

The result was a deeply compromised and vaguely worded Framework Convention on Climate Change. A target was proposed for stabilizing greenhouse gas emissions from wealthy countries; but no country was required to meet this goal, few penalties were discussed and none was put in place. The US administration thoroughly alienated world opinion and left the conference isolated from all but a few countries. European countries, Japan, and Canada expressed a readiness to accept binding commitments if the US agreed to do so. However, US negotiators, and the ever-present and powerful fossil fuel lobby, were sharply opposed to such a move. In fact, the first President Bush made his appearance at the conference contingent on the rejection of any concrete commitments. It was quickly evident that this agreement would have no impact on actual emissions; but some argued that a weak treaty with the US on board was better than a stronger treaty that did not include the US. Many came to regret that conclusion.

Responsibilities were allocated by groups of countries: wealthy countries, former countries of the Soviet bloc, and underdeveloped countries. While recognizing the responsibility of all countries, the treaty asked wealthy countries to take the lead in addressing the problem. It also called on countries to adopt the precautionary principle, not await conclusive proof of a major crisis before taking action. And it stated that the extent to which impoverished countries were expected to implement their commitments was dependent on the effective implementation by wealthy countries of theirs. The idea was that early reductions by wealthy countries would lead the way; and technology and assistance would be made available to impoverished countries to help them reduce their emissions in the future.

The US was instrumental in ensuring the ambiguity of the final agreement. Impoverished countries pushed for a specific principle to be laid out that might guide future negotiations. The US rejected not only clear targets but also the delineation of any specific principles. Indeed, the very word 'principles' was removed from the wording of the Earth Summit Convention.[60] President Bush was virtually alone among the leaders of the world's wealthiest countries in his refusal to accept strong policies to address climate change. For many, when coupled with the previously mentioned US rejection of the Convention on Biodiversity, this signaled a gross American disregard for protection of the global ecosystem.

Even watered-down half-measures faced an immediate backlash. Although there was strong potential to adopt more efficient practices and reduce waste to meet early reduction goals, oil industry leaders, energy companies, and trade groups cited inflated cost figures and promised that any required reduction would lead to economic slowdown and job losses. A corporate assault on the conclusions of the Intergovernmental Panel on Climate Change was launched, including public pronouncements full of foreboding and pseudo-scientific reports paid for by major energy and oil interests.

As mentioned previously, when President Clinton took office in 1993, he declared his support for international initiatives to address climate change and announced a national effort to meet European targets and stabilize domestic greenhouse gas emissions at 1990 levels by 2000. Clinton and Gore had campaigned hard against the Bush administration's performance at Rio, and American environmental leaders seemed to have reason for hope. However, as we saw with domestic policy, there was a big gap between rhetoric and reality. After a fair bit of presidential waffling and Clinton's failed effort to enact an energy tax, optimism faded. Strong corporate lobbying and resistance from key lawmakers stymied progress. Voluntary measures were proposed, including the use of energy-efficient lightbulbs and building materials – hardly sufficient to curb the trend toward global catastrophe. In the end, a drop in gas prices and the rising popularity of gas-guzzling SUVs eliminated any gains from the so-called Climate Change Action Plan. Greenhouse gas emissions were greater than ever in the US.

By the time of the first meeting of the Conference of Parties of the Framework Convention on Climate Change (FCCC) Berlin in 1995, only the UK and Germany were on target to return their emissions to 1990 levels. Somewhat prior to the meeting, the IPCC announced that recent research indicated that even a roll-back to 1990 levels would not be enough to prevent human-caused global climate change. To appease US demands, several measures were proposed to attempt to relieve the pressure on corporations in wealthy countries. The idea of so-called joint implementations would allow developed countries to undertake projects in the impoverished world to reduce emissions instead of reducing their emissions at home. US negotiators claimed that this would allow for more 'economically efficient' means of reducing global emissions. Impoverished countries opposed this idea, fearing that this was yet another effort to transfer the burden to poor countries.

The most important result of the Berlin meeting was a schedule to adopt specific emissions-reduction targets. These targets would be applied only to wealthy countries, or so-called 'Annex I' countries. European countries and most developing countries supported these clear targets. Some oil-producing countries, and a few impoverished countries that feared such limits might eventually be placed on them, opposed such measures. Small island nations, already feeling the early effects of rising ocean levels and severe weather, proposed a reduction of greenhouse gas emission by 20 per cent by 2005, and gained the support of several environmental groups. India came out in support of the proposal, prompting other impoverished countries to follow suit and defining the birth of the so-called 'green G77'. Some oil-producing countries, fearful of being marginalized in the G77, supported the measure as well. With the EU and G77 now in support of clear targets, the pressure on the US was sharp. While the Senate Republican majority remained opposed to such targets, the Clinton administration accepted this so-called 'Berlin mandate,' committing the US for the first time to some form of reduction.

At the same time the US began to push for tradable pollution credits. US delegates made the dubious claim that the US had achieved success in reducing sulfur emissions with such policies. Countries that had accepted specific emissions targets would be granted the right to a certain volume of emissions in the form of a credit. If they emit less than they are allotted, they could sell their remaining credits to other countries. This has the perverse effect of rewarding the greatest polluters of the past by giving them valuable credits. Impoverished countries, since they would not have specific reduction targets, would not be allocated credits. As counties in the impoverished world quickly realized, this could allow a country like the US to meet its obligations by buying credits without actually reducing its emissions. Since countries like Ukraine and Russia would be allocated credits based on their past emissions, and the slowdown of their economies meant they would be likely to have sharply reduced emissions with or without restrictions, there would be plenty of credits for sale. Former Soviet countries could cash in on the emissions credit system while countries in Latin America, Asia, or Africa could not; and the US could meet its obligations without actually reducing its emissions. This struck many as extremely unfair. The fact that impoverished countries had not polluted greatly in the past would seem to work against them. Together with measures such as joint implementations, the US seemed to be pushing for a system that interpreted reduction targets as the right to pollute and that locked in existing inequalities in the use of natural resources while creating a lucrative market for corporations in the buying and selling of pollution!

At the same time, opposition within the US Congress was growing more determined. In particular, US lawmakers objected to the fact that only wealthy countries would have to meet clear targets in the first phase of reductions. In June 1997, US senators Robert Byrd of West Virginia and Chuck Hagel of Nebraska introduced a resolution that called on the US to reject any binding targets if specific reduction requirements are not also applied to developing countries or if such reductions would cause 'serious harm to the economy of the United States.' Senator Hagel went on to argue that the current agreement would threaten US economic growth and military security (the military is a major user of fossil fuels). The measure passed 95:0. Meanwhile, WWF rated the US last among the world's twenty major wealthy countries in efforts to reduce carbon dioxide emissions.

Critics claimed that any restriction on US emissions that did not also restrict the emissions of less developed countries would be unjust and environmentally ineffective. This was deceptive at best. While targets were being discussed only for wealthy countries, all countries were obliged under the general climate convention to take action. Hence no country would be off the hook. More crucially, the emissions from impoverished countries were far below those from industrialized countries; and even the fastest growing countries would not catch up for several decades. US production put about ten times the amount of carbon dioxide into the atmosphere as China, the largest producer of greenhouse gases in the impoverished world. When considered per person, the inequity is even greater. As we've seen, the average American produces many times the average greenhouse gas as the average person in China. But this wasn't really the issue anyway. Some of the opponents of the agreement may have genuinely wanted greater commitments from developing countries; but for most opponents, the insistence on specific reduction targets for impoverished countries provided a convenient justification for squashing movement toward an international agreement.

Industry associations funded a $23 million television campaign portraying the 'injustice' of any agreement that did not include developing countries and warning that fuel costs would soar and US economic competitiveness would be severely harmed. The Global Climate Change Coalition, a consortium of coal, oil, and car companies, fought hard against binding limits and used considerable political leverage to get their voices heard. The Coalition for Vehicle Choice, a group financed by the auto industry, headed a major effort to convince civic and small business groups that limiting emissions would be bad for America. Industry groups and their paid scientists attacked the science of climate change, arguing that the scientific conclusions were not sufficiently certain to warrant the economic cost of restrictions. Grossly inflated estimates of the cost of emission reductions were widely publicized.

In this context, in December 1997, the third conference of the Climate Convention took place in Kyoto, Japan. The European Union representatives argued for a binding commitment toward a 15 per cent reduction of carbon dioxide, methane, and nitrous oxide by 2010 for all wealthy countries. The US, on the other hand, had declared that it was not willing to go beyond stabilization. However, US vice-president Al Gore promised greater flexibility on further reduction targets if the final agreement included language that would open the door to reduction obligations on the part of impoverished countries. Feeling pressure from industry groups and their political allies, and acutely aware of the demands of the Byrd–Hagel resolution, the Clinton administration saw some form of commitment on the part of impoverished countries as necessary for any agreement to get through Congress. But since the agreement in Berlin specifically excluded impoverished countries from definite reduction obligations, this would be nearly impossible.

With the US as the largest single polluter and the world's largest economy, its concerns defined the conference. Negotiations were contentious and dragged on beyond the deadline. The final agreement showed all the scars of hard-fought compromise. After listing a menu of suggested measures to improve efficiency, reforestation, renewable energies, and sustainable agricultural practices, the agreement outlined reduction targets. The targets were modest, but they did call for specific reductions and therefore defined what could be a valuable first step. The US would agree to a 7 per cent reduction from 1990 levels. The fifteen nations of the European Union would reduce emissions by 8 per cent. But since European Union emissions would be considered as a whole, some European countries would be required to reduce their emission more than others. Canada and Japan would reduce their emissions by 6 per cent. In line with US demands, the Clean Development Mechanism allowed countries to invest in sustainable development projects in the impoverished world in exchange for emissions credits. European proposals for required policies and measures to reduce emissions were, on US insistence, instead defined as recommendations.

Negotiations in Kyoto went through the night. US attempts to open the door to specific targets for developing countries were vigorously resisted by China and others, who reminded the delegates of their existing obligations under the Berlin agreement. As dawn approached, an agreement was reached, but Article 9, setting out terms for participation by developing countries, had been deleted and provisions for carbon

6.9 / Kyoto commitments: initial CO_2 emissions reduction commitments under the Kyoto Protocol (% of 1990 levels)

European Union	–8
United States	–7
Canada	–6
Japan	–6
Australia	+8
Iceland	+10

trading were left vague.[61] So, despite the US negotiator's clear success in redefining several aspects of the agreement, the Kyoto Protocol did not include binding targets for developing countries and thus did not meet the nearly impossible demands of the Byrd–Hagel resolution. The Clinton administration nevertheless signed the agreement.

Criticism of the agreement from the administration's opponents was instantaneous. Such a treaty would 'destroy the US economy,' leading Republicans argued. House Speaker Newt Gingrich led the charge, calling the proposed treaty 'an outrage,' and calling the administration's motives 'environmental correctness and inconclusive science.' Senator Chuck Hagel declared, 'I want the president to send this agreement to the Senate. We will kill it.'[62]

The strategy of corporate lobbyists and industry pressure groups worked as planned. Oil, gas, automobile, and coal industries – the so-called carbon club – poured millions of dollars into the efforts to squash the Kyoto Agreement. They had successfully driven a wedge between wealthy and impoverished countries by whipping up demands in the US that impoverished countries must agree to binding reductions, despite the fact that the Berlin agreement specifically excluded this in the first round of reductions. Months of intense lobbying, dubious reports citing questionable research designed to scare American workers, and a well-funded public relations blitz on television and radio had effectively moved the goalposts. The American Petroleum Institute worked aggressively to sway public opinion and undercut the scientific consensus on human-induced climate change. The industry group paid for dubious research that warned of job losses and wage decreases if the treaty were ratified. Organized labor, with the fear of job losses drilled into them, fell into line and voiced opposition as well. House Democratic leader Richard Gephardt, a long-time labor ally, declined to support the administration's efforts. For industry, global warming wasn't an environmental crisis, it was a public relations problem.[63] They just needed to convince the public that global warming wasn't a problem. Ignorance is bliss.

US support for the treaty was crucial to its survival. The treaty would not come into force until at least fifty-five countries that contribute at least a combined 55 per cent of global greenhouse gas emissions ratify the agreement. Since the US accounted for more than a third of emissions from wealthy countries, enactment of the treaty would be difficult without US ratification; and some countries including Japan made it clear that they would not ratify the agreement unless the US, as the world's largest polluter, did so. Indeed, with US per capita emissions at 20 tonnes per year, five times the global average, there was good reason to demand that the US lead the way. The ratification process stumbled along; but, without the ratification of the US, the prospects for the Kyoto Agreement seemed dim.

Faced with industry resistance and the intransigence of key senators, pressure on the Clinton administration to secure participation from the impoverished world increased. Ratification from the Senate did not seem possible without it. Congressional opponents rejected a Clinton proposal that would require developing countries to accept binding emission targets by 2005. In response to US urgings, Argentina promised to take on voluntary commitments. An attempt to get other developing countries to adopt voluntary commitments at the fourth conference of parties in Buenos Aires was completely rejected by the G77. India, supported by China, argued for a per capita quota on emissions. In other words, a country's quota would be based on its share of the global population. Desperate to get some form of commitment from China, India, and the G77, the US did not object. But, with each American consuming thirty times the energy of the average Indian, defining such targets was left to some future negotiations.

To take the pressure off the US, the Clinton administration proposed at the end of 2000 that all 'managed lands' be considered 'carbon sinks' and counted as credits towards a country's reduction commitments. Since almost all land in the US is managed in one form or another, as forests, plantations, or farms, this would tremendously reduce US obligations. This offered a giant loophole, allowing the US to continue high emissions by taking credit for actions it would have taken anyway. Moreover, critics argued, forests function more as a storehouse for carbon rather than a continuous sink; the persistent capacity of forest to absorb carbon dioxide is not clear, making reductions a far more certain and safer option. The

European Union and members of the G77 opposed this proposal as an obvious effort by the US to reduce its obligations. Negotiations in the Hague in 2000 revealed the depth of the differences between parties. After days of fruitless debate, the talks collapsed.

When talks reconvened in Bonn eight months later, President George W. Bush had taken office. In a reversal of his campaign agenda, President Bush immediately rejected the Kyoto Accord as 'fatally flawed'. Conservative pressure groups urged the administration to 'unsign' the treaty so that the question of ratification would not arise.[64] Europeans tried to dissuade the new American leadership, to no avail. Rejecting the overwhelming consensus among scientists, Bush clung to untenable industry claims that the scientific evidence on climate change was inconclusive. Even the Union of Concerned Scientists found that the administration repeatedly suppressed and distorted scientific research to suit its political aims.[65] President Bush refused to recognize a report from his own EPA that acknowledged the human causes of global climate change. When a State Department report to the UN acknowledged the clear human causes of climate change, President Bush dismissed it as a 'report put out by the bureaucracy.'[66] The administration manipulated scientific research and tampered with the findings of federal agencies to make a series of EPA reports appear to support the Bush administration's deeply flawed position.[67] Whatever the cost to the planet, the Bush team seemed determined to stick with its benefactors in industry.

When negotiations resumed in Bonn in July 2001, the potential for a breakthrough seemed bleak. However, without the need to appease US industries and policymakers, a compromised deal was reached in several issues. It would be agreed that funds would be established to help impoverished countries, particularly the most direly impoverished, gain access to technologies, implement sustainable measures, and adapt to the consequences of climate change. Sinks such as forests and croplands would count toward a country's emissions reductions. Critics pointed out that the deeply compromised reduction requirements could be met with relatively small increases in energy efficiency, a limited emissions trading scheme, and perhaps a small carbon tax. Maintaining some forward movement was seen as crucial. It was hoped that the US might join later.

By the start of 2005, the Kyoto Protocol was dying a death of a thousand cuts. The European Union had become the treaty's major promoter, but could not muster the necessary courage to meet its reduction commitments. The Bush administration never missed an opportunity to reiterate its opposition to the treaty and continued skepticism about global climate change. Russian president Vladimir Putin, enjoying the fruits of tremendous economic expansion, grudgingly signed on to the protocol in exchange for Europe's support for his bid for WTO membership. But with the US out of the agreement and global emissions continuing to grow, it's clear that this accord has become at best inadequate, and at worst a farce. Meanwhile, ocean levels are rising, species are dying, glaciers are melting, diseases are expanding; forests, coral reefs, and other precious ecosystems around the world grow increasingly endangered. Like the infamous Nero who fiddled while Rome burned, our leaders do nothing but fiddle; but this time it's the world that's burning.

Resistance is fertile / 7

The previous six chapters have made the reality of the global system all too clear. We're raping the earth, destroying the precious and fragile natural systems that support life on this planet, casting men, women, children, and the precious gifts of nature alike as fodder in this rapacious assault, and ultimately degrading our very humanity. Whether we like it or not, whether we want to believe it or not, our current way of doing business is simply physically unsustainable. More than this, to anyone with a conscience, with a soul, the workings of this global system must be found reprehensible. Efforts to lessen the destruction, to preserve some of our most vital natural resources or moderate the worst excesses of an unrestrained market, have been hobbled by corporate power and political surrender to a free-market dogma. You're outraged. You're angered by the capitulation of your elected leaders to corporate privilege; you're offended by the inhumanity of this system; you're alarmed over the mounting environmental crisis. Well, now it's time to do something about it.

There's no doubt that the present global order is powerful and deeply entrenched. But not a thing about it is inevitable, and nothing makes it immune to demands for change. This system was built by human beings. The institutions that define and enforce the rules are run by human beings. The free-market fixation on profit that appears to justify this system was invented by human beings. If enough human beings want change, then we will have it. A global transformation is more than possible; it is necessary, and, if we are to survive as a species, it is inevitable.

Buy a hybrid car, save the world?[1]

So, you're ready to take action. You just need to know what to do. What letter should you write and to whom? What environmental group should you send your check to? Where should you shop? You'll promise to recycle more, conserve water, try to drive less, and even take the train when it's convenient. A litany of 'how to' books will give you other ideas. There's *The Recycler's Handbook*, *50 Things You Can Do to Save the Earth*, or *Save Our Planet* (which offers a full 750 things you can do to save the planet). Or how about *You Can Prevent Climate Change*? Each of these offers an upbeat list of changes that would make your home, office, and school more environmentally gentle. You're ready. 'Just give me the list,' you're thinking, 'I'll do 'em all.'

Sorry, it's not that easy. We are not going to find salvation in lifestyle changes. Riding a bike to work, recycling, carrying a reusable shopping bag, or planting a tree will not make the current global system either morally or physically sustainable. The focus on these sorts of lifestyle solutions reveals what I see as a grave weakness in the dominant environmental agenda. We've become passive, accepting, and far too timid. We're trying to fix a machine that is fundamentally broken by tinkering lightly on the edges. Most disturbing of all, we seem to have accepted, in a matter-of-fact sort of way, the legitimacy of the individualistic, profit-centered thinking that underpins the system we're trying to change. As a result, we imagine environmental solutions as consumer-based, and friendly to the consumption-centered world-view that inflicts the global crisis.[2]

As a result, too many of us have come to see the environmental crisis as our fault, as the result of individual shortcomings and self-indulgence. We blame consumers in wealthy countries for buying too much stuff. But we don't talk about the omnipresent marketing that defines these manufactured needs and whips consumers into a perpetual shopping frenzy. We blame impoverished people around the world for having too many babies. We don't address the desperate and disempowering conditions that lead to rapid population growth. Instead of attacking the coercive power infrastructure that enforces the current order and constrains our options, we're left pointing fingers and dodging guilt. Who's responsible for global warming? Not me in my hybrid, it's that guy in his SUV!

The great advantage of this perspective for the corporate order is that it puts the burden of change on individual consumers. Corporations justify the lethal by-products of their industry by pointing out that people purchase the final merchandise. So we're led to believe it's OK to make anything in any way just as long as you can sell it. If you're an active environmentalist, you've heard this line of reasoning before – 'We're just making what people want, and people want Hummers.' Or as the timber industry likes to say, 'We'll stop cutting down forests when people stop buying paper and wood products.' This line not only allows them to deny culpability; it also ensures the security of the status quo. Consumers are unlikely to miraculously and *en masse* step out of the coercive network of marketing and commodification that defines their world and dramatically throw down the chains of consumer culture. And, of course, corporations work to make sure that this doesn't happen. They're careful to hide the real costs of their products with multiple coats of greenwash and they staunchly oppose any labeling that might allow for more informed consumer choices. They know that their profits and their privileged position are secure as long as the search for solutions remains fixed on consumer-based options. In short, we've let corporate America define the terms of our resistance. We're left trying to convince people to drive a little less or recycle a little more, hoping that such lifestyle changes might save us; but they won't.

The truth is that it's simply wrong to allow people to believe that they are personally responsible for the global ecological crisis. It shifts our attention away from the real problem. It lets the corporate order off the hook, and it ignores the entrenched power structure that constrains our alternatives. By focusing on individual choices we atomize our resistance and undercut our capacity for solidarity. Instead of a movement in demand of systemic change, we're a bunch of consumers trying desperately to make the right purchases to ease our tiny share of the global burden. In the end, of course, we fail to find a balm for our green guilt. And so we are less likely to make a fuss about corporate sins. Instead of demanding change from those who are most culpable, we repeatedly rehearse the notion that proper consumer choices can save us. The reality is, they can't.

The real options – the ones that could save us – are not a matter of individual consumer choice. While our commodified lives offer thousands of daily selections, we are left with increasingly few meaningful alternatives. You can pick from a hundred brands of toothpaste, each with minute variations in style and marketing. You can decide to shop at any number of faceless big package stores. You can pick from a variety of tasteless food items in any number of identikit restaurants. You can even choose a green option... well, *sometimes*. But, in the end, we're left with fewer and fewer real alternatives. You can choose between dozens of different car models, with hundreds of variations in style and color, each one modified every year to enhance your consuming pleasure; but you can't choose to have an adequate public transport system that is sufficiently supported by public funds. You can choose between countless brands of bottled water, but you can't choose to have clear rivers and lakes and safe water from your tap. At least you can't choose these things on your own. You can't decide as an individual to live in a sustainable society. As long as your options for action are cast in individual consumer terms, real change, the kind that is necessary to save us, will not happen.

This may be tough to hear. The global crisis is scary stuff. What are we supposed to do when faced with global warming, ozone depletion, mass extinction, and widespread human suffering? Not knowing what to do with our concern, we feed our kids Save the Rainforest breakfast cereal, we make sure our tuna is dolphin-safe, and we order that hemp shopping bag from the environmental group's website – that'll show corporate America! Desperate to make a difference, we do what we think we *can* rather than what we know we *must*. The effect can be perverse. We drive across town to buy fair-trade, eco-friendly coffee. We run the dishwasher to clean glass jars for recycling (a process that consumes far more energy than the small amount that's saved by recycling).[3] We can lament global warming and rail against high gas prices in the very same conversation. Our dissent is commodified and sold back to us so we can purchase our green credentials (if we're rich enough to buy organic food and a hybrid car). An array of eco-friendly goods promise that salvation is achievable; we just need to buy more earth-friendly products. In the end, our pangs of green guilt are eased a bit. But the truth is we can't consume our way to sustainability and justice.

Please do not misunderstand me: there's nothing wrong with doing your best to minimize your personal impact on the earth. So I am *not* saying that these lifestyle changes aren't worthwhile. I'll admit right here to being a composting, recycling, bike-riding, vegetarian. If composting and driving less destructive cars helps postpone some of the destructive effects of the global crisis long enough to build a movement for real change, that helps. But we need to understand that is what we're doing, buying time. We need to recognize that without the simultaneous struggle for systemic changes, lifestyle changes may mean very little. At worst, if they give us the false illusion that our current course is sustainable (that car-centered culture is OK so long as we all drive hybrids), then these measures can undercut the pressure for more crucial systemic change.

And I'll readily agree that our efforts to tread lightly on this planet can enrich the struggle for systemic change. From recycling to riding a bike, these choices can help confirm, publicize, and reinforce our dedication to the earth and each other. Recycling may not save the world, it may not even save much energy, but if it reconfirms your commitment to the struggle for change, it's worth doing for that reason alone. Carrying a shopping bag may only save a few hundred plastic bags in your lifetime, hardly enough to change the course of our global crisis, but if it provokes discussion with your grocer, if it allows you more readily to recognize your own consumption, then it becomes an integral part of the struggle. But let's be straight: not being part of the problem doesn't make you part of the solution.

Unfortunately the leaderships of mainstream environmental organizations in the US and in other wealthy countries have not led the way. Too many of them have grown satisfied with demands for minor modifications that *might* slow down *some* of the most destructive processes of this global suicide machine. They seem to have given up hope for more systemic change. But I don't mean to cast stones, at least not big, heavy ones. I understand the pressures they face. An organization that is continuously ringing the alarm bell, declaring an environmental emergency, and demanding radical change is going to have a tough time holding on to its donors in suburbia. To survive, environmental groups have to look successful while carefully avoiding collisions with real power.[4] Pushing lifestyle changes and minor reforms allows them to present the sort of upbeat list of environmental solutions that make Middle America happy, allowing them to believe they can be environmentalists without surrendering their privilege or dramatically changing their lifestyle. Facing an entrenched corporate order, established environmental groups have adopted a strategy of compromise, market-based policies, and corporate-friendly reforms. Unfortunately, in the real world, in almost every case, these strategies are failing. And, once again, if these compromised policies allow us to feel that the earth is adequately protected, if they mask the depth of the ongoing crisis, if they distract our attention from the struggle for systemic change, they do more harm than good.

We need to do better than this. If we are to survive as a species, we need to rethink the very structure of the global order. We need to start thinking about real options rather than minor alterations to an unsustainable system. We need to recapture our environmental imagination, our collective ability to envision alternatives to this consumption-centered, corporate-controlled world. This means rejecting the taken-for-granted myths that perpetual growth is indispensable to our happiness, that the market will provide all things in the best way possible, and that deep compromises in our health and well-being are a necessary price for continued economic development. We need to abandon the idea that a system fixated on profit – and only profit – can produce a socially just and sustainable society. We are not just consumers. Society is more than a bunch of self-absorbed shopping addicts working toward their next fix. If it's not, then it damn well should be, and it can be. We need to recognize that consumption is not happiness and happiness is not consumption.

In this vein, many argue, the most urgent task of environmentalists may in fact be to foster a sort of cultural renewal, a rejection of prepackaged, consumerist society. Countless environmental activists, researchers, and writers have noted that ecological dislocation and social exploitation are inherent to a culture of limitless materialism.[5] So, until we address our own narcissistic, commodified socialization, both environmental sustainability and social justice will remain out of reach. A hundred and fifty years ago Henry David Thoreau had already recognized the 'spiritual poverty' of unlimited greed and an alienation from nature. More than a century later, it is clear that Western nations have slipped deeply into a culture of contentment, pacified by consumer goods but never really happy. But this fixation on immediate synthetic comforts at the cost of long-term happiness did not develop organically. As we have seen, it was shaped and imposed by power. At its base, the environmental crisis and deep injustice of the current order are not really about culture, not about science, technology, or planning. They are about power. It is in the end a political problem that demands a political solution.

Challenging the rules of the game

Some would have us believe that collective demands for policy changes are wrong-headed and destined for failure. Corporate powers and their supporters in government would like to depoliticize the environmental crisis and keep the market outside accountability. They'd like you to believe that your government should not and cannot be responsible for protecting the environment. For policymakers, this argument provides a convenient claim of impotency. If the market alone can provide sustainable options, public servants need not risk the wrath of the corporate order by pushing for significant environmental protections. We hear this from our leaders all the time. 'Good conservation and good stewardship,' President Bush recently assured Americans, 'will happen when people say, "I'm not going to rely upon the government to be the solution to the problem."'[6]

Advocates of this retreat from public control and accountability cite the famous example of the 'tragedy of the commons' to make their point.[7] The story goes something like this: an area of pasture, 'the commons,' is shared by all the villagers. The problem is that each member of the village benefits by allowing more of his or her sheep to graze on the commons. Since choosing not to graze sheep on the commons would not stop others from overgrazing, there is no incentive to restrict one's use of the common land. As a result, in the end the land is degraded by everyone's overgrazing, leading to hardship for the entire village. If this land were privately owned instead of publicly shared, the storyteller often concludes, the commons would not have been overgrazed, since each owner would have had a private interest in preserving the land's productivity.

The story is interesting; the conclusion is ludicrous. A quick look at the track record of private corporations with the stewardship of nature makes it immediately clear that the private ownership of nature can often have tragic consequences. Yet the hard-core free-market advocates continue to push their point. The Bush administration has even proposed that we start privatizing our national parks. Anyone who would suggest such a thing simply does not understand the meaning of stewardship. Nor can privatized solutions begin to address the most pressing aspects of the current crisis. Should we also privatize the atmosphere? The oceans? The ozone layer?

The free-market loyalists would have us forget what is patently obvious: it is the government's job to protect our shared and vital interests. That's why governments exist. And no interest is more shared and more vital than the preservation of a livable planet. Governments are supposed to help us resolve the tragedy of the commons scenario (or what social scientists call collective action problems). We know that cooperation, or 'collective action,' is sometimes difficult. Even though cooperation within a group could lead to a better outcome for everyone, often members don't cooperate because they don't trust others to do so. So, for example, we might all be better off if we were to stop driving SUVs. But since I don't believe that others will stop driving theirs just because I stop driving mine, I'll keep driving. Rather than take on the burden of cooperation (by driving less) and having my efforts undercut by those who don't take part in the agreement, I choose to opt out early. This is a problem. (And it's fundamentally the reason consumer-based environmental solutions alone won't save us.) But the whole point of government is to address this problem, to provide an institution that can protect our shared interests and ensure that everyone cooperates with the rules we collectively make to protect our common welfare. Governments can and should allow us to make rules about the kind of cars we drive or the ways we will care for our most vital resources.

The lesson from the tragedy of the commons story is not that natural resources must be privately owned to be protected. Instead, this story, and the past several centuries of our history, should tell us that unmanaged and unprotected natural resources are vulnerable to abuse from the greedy.[8] The solution is to define durable agreements, practices, and rules that ensure a shared stewardship of nature. Sometimes public institutions are necessary to define or enforce these rules. Sometimes this can be done cooperatively, without strong government regulation.[9]

Put more directly, the assumptions of the tragedy-of-the-commons story (the same assumptions that underscore neoclassical economic theory) are simply wrong. Anyone who's been active in the struggle for social and environmental justice can think of many instances where community members have collaborated rather than competed to solve shared problems. The presumption that what is good for the community is bad for the individual and that self-centered, antisocial decision-making is the norm, are challenged by a wellspring of communities and individuals around the globe who are putting their shared interest and the earth first, and finding ample rewards for their efforts. Noted authors David Suzuki and Holly Dressel have uncovered a multitude of examples that give the lie to the presumed conflict between saving the earth and enriching our lives.[10] Ranchers are learning that preserving rangelands and habitat for endangered species can also mean a better product and a more rewarding job. Around the world small farmers are rejecting the mass-production methods of industry and being rewarded with healthier lives and more productive soil. Urban designers are imagining new design concepts that enhance our lives and health, build stronger communities, and conserve natural resources in the process. Green entrepreneurs are defining new business models that don't just accept but truly embrace nature's limits and bear lightly on the planet. Learning from natural systems, local businesses and governance are defining a new vision of industrial ecology that is rethinking traditional production practices.[11] These may not save the world, but they are clearly steps in the right direction. More importantly they remind us that social and environmental decision-making and long-term thinking are not only possible, they can even mean short-term benefits, possibilities excluded by the tragedy-of-the-commons scenario.

There are thousands of small success stories out there, from organic farmers to fair-trade collectives, which remind us that change is possible. We do not need to wait for scientific breakthroughs to solve these problems; we know right now how to save the earth and build a more just society.

It is time we started redefining the rules of the game to encourage these alternatives and protect the things that we say matter.

If we view the world through the lens of neoclassical economic dogma, it is easy to imagine ourselves trapped in an endless collective action problem – I won't change because no one else will, and they won't change as long as I don't. If we instead begin to imagine our alternatives as societal changes rather than individual choices, it is easy to see that this need not be the case. In reality, each of us regularly puts aside our own immediate pleasures for the sake of a commitment to others. We are fundamentally social beings. We hug our children, help our neighbor, participate in the PTA, rally in the streets, pick up litter, and we vote. This last item alone is enough to reject the dogma of self-centered individualism implicit in free-market economic theory and the tragedy-of-the-commons story. If we were truly as self-seeking and asocial as economic theory seems to assume, no one would ever vote. After all, voting can be burdensome, and you can be fairly sure that your one vote will never change the outcome of an election. But that's not why we do it. We vote because we feel it is our duty, part of our deep commitment to our greater community. We have learned that voting is important, and we do it quite simply because it matters to us. In this and countless other ways, we act like committed members of a community every day. In short, we're not as self-centered as economic theories would imply. But, you're thinking, fewer of us vote with every passing election. Isn't this a bad omen? Yes it is. And it's time we changed that.

Taking back democracy

Near the end of his classic cautionary tale, Dr Seuss's Lorax tells us 'UNLESS someone like you cares a whole awful lot, nothing is going to get better. It's not.' But to achieve real change we need to do more than care, or even care *a whole awful lot*; we need to make others care. We need to make those in power care. We need to catalyze a sea change in the way we understand our society, our government, our economy, and our planet and we need to make sure that those in power recognize this change. Politicians don't want to talk about real change; it can put them at odds with their corporate sponsors and divide political parties. So we can expect them to dismiss our demands as unreasonable or too extreme. They'll remind us that environmental protections must allow for the continuation of growth. They'll recite the orthodoxies of market economics, and tell us that an end to poverty can only come through continued corporate expansion. Deeply beholden to the corporate order, they will desperately avoid discussions of consumption, expansion, corporate control, and real justice. They will seek the last refuge of scoundrels, and inform us that it is our patriotic duty to consume without question. The reality is, of course, quite the opposite.

Achieving the ability to demand changes requires that we organize and build political muscle. This won't be easy. Too many people live in denial. Too many feel helpless. Too many can't identify real alternatives through the pernicious haze of corporate-induced ignorance. We can change this. A good first step might be rejecting the dichotomy that casts consumer culture as fun, fulfilling, and patriotic, and working for change as self-sacrificing, quixotic, and perilous. Contrary to popular perception, resisting consumer culture doesn't require donning a hair shirt, becoming a martyr to the cause, or giving up your dearest indulgences for a monastic lifestyle. It's increasingly clear to me, and needs to be made clear to many others, that placing your own life in greater harmony with the earth and working for social change can reap 'double dividends.'[12] It can both enrich our everyday lives and preserve our planet. We can build meaningful alternatives to the cultural and economic binds that commodify our hopes and desires. We can build healthy and livable neighborhoods, create clean jobs, foster richer communities, uncover more meaningful lives, fight disease and alienation, and even enjoy better food as we build a more survivable and just world. And herein lies our strength. People may imagine their self-interest too narrowly to include deforestation, but certainly protecting one's children, finding a more meaningful life, and eating and breathing safely matter to even the most unenlightened.

There's no secret formula to building a movement for social change. We've done it successfully in the past on several occasions. Workers demanded and won minimum wage laws, an end to child labor, and other protections from the excesses of the free market. Government leaders answered their initial demands by claiming that they had no power to regulate the labor market. They cited economic theory and warned that regulating such things might destabilize the economy. They accused them of lacking patriotism. But faced with massive protests that undercut their claim to represent 'the people,' political leaders had no choice but to concede reforms. All great changes have come from below. A woman's right to vote, civil rights, and protection for workers came in response to demands not requests. They came in reaction to a groundswell of informed and organized resistance. All of these movements seized the moral high ground, exposed the hypocrisy of the system, threatened the legitimacy of those in power, and insisted that our policymakers live up to our highest principles.

Many argue that rallying civic action toward change has gotten tougher in the past few decades. Social research confirms what is already obvious: Middle America has grown increasingly disparate and alienated. Town centers have disappeared, replaced by culturally sterile, homogenized malls. The community groups and civic associations that once defined a sense of shared understanding seem to be things of the past.[13] Town cinemas have been replaced with DVDs. Gated clusters of socially barren McMansions have displaced neighborhoods. The commodification of society is increasingly turning America into what the market would have it be: apolitical, atomized, self-serving consumers. With this sense of community goes a sense of responsibility and efficacy. Couple this with the increasing marginalization and impoverishment of countless communities and you're left with too many citizens thinking that they can't change things.

Yet this trend is countered with a far more encouraging development. Across the country, a diverse array of citizen groups and community activists are mobilizing for real social change. There is a clear rise in innovative, citizen-led, collaborative democratic action. Local efforts to build safe and healthy communities, civic environmental groups, community efforts for healthcare, citizen environmental monitoring projects, community education programs, cooperative efforts to fight homelessness and many other examples all signal the potential for a new civic engagement. They aren't always explicitly environmentalist, but they are increasingly coming to recognize the environment as an integral part of their struggle. Often disappointed with the failure of traditional public institutions or established organizations to address their problems, community boards, school groups, neighborhood associations and others have turned to a locally focused, self-help model of civic participation. They don't have boards of directors, offices in Washington DC, or national staffs. Instead they are often deeply local, generally informal, and usually inclined toward democratic decision-making. They have come to believe that a movement is most powerful when its members are most empowered, and they are defining a new mode of civic engagement that will do just that.

As these movements demand a voice in the forces that shape their lives, they should give us hope. Not just hope that we can influence a few policy decisions, but hope that we can take back control, real control. We need to do more than take back political control; we need to redefine the bounds of the political, re-imagine democracy, and in the end democratize the economy.[14] That is to say, we need to ensure that market forces have democratic limits. We know that corporate power can impede democracy, squashing democratic voices, stifling efforts to protect our shared interests, and seriously constraining the limits of public debate. Private interests do not have unlimited rights to do as they please simply because they're rich. We need to decide which matters most, corporate power or democracy. To the workers, parents, students, and neighbors mobilizing around the country, the answer is clear. To them it seems self-evident that any right held by a corporation is secondary to the right of citizens for a safe place to live, a meaningful future, and accountable governance.

This is not a new demand; it is the fulfillment of an age-old promise. After all, if we have a right to 'life, liberty and the pursuit of happiness,' then this must entail a right not to be poisoned. Even if we have only a right to life, a safe environment in which we can live and the preservation of a livable planet must be equally inalienable. If the Constitution protects our privacy and personal integrity, how can it not also protect us against the trespass of toxic chemicals into our homes and communities? If the fourteenth amendment ensures equal treatment under the law, then surely unequal protection from exposure to toxic waste, poisoned air, and lethal water, based on the color of one's skin or the size of one's pocketbook, must be prohibited.[15] It's time that we start demanding the full breadth and meaning of our highest ideals.

It is time we took back our democracy. Government has become increasingly distant, estranged from our everyday lives. We talk about it as though it's a thing, out there, far away from us. But it's not; it's our government. But our representatives, the protectors of our vital interests (our employees!) have become remote, alienating, and intimidating. This is hardly the institution you want protecting the planet or your health. But just because that's the way things are, doesn't mean that's the way they must be. Democratically accountable public institutions can and must serve as the counterbalance to powerful corporate influence. It's time to ensure that our elected representatives act in our interest, not in the interest of their corporate donors.

The new environmentalism

There is a new environmentalism taking shape, and it isn't just multifaceted; it's multiracial, multigenerational, multiclass, and multicultural. This new environmentalism merges currents of the civil rights movement, community antitoxins activism, the labor movement, and Native American struggles, to name just a few.[16] From church groups around the country declaring 'God is Green' or bitingly asking 'What would Jesus drive?' to urban immigrant neighborhoods rallying for safe green spaces for their children, there is a new face to environmentalism. Vulnerable communities of color are standing up to environmental injustice and asserting their rights to safe working conditions. Indigenous peoples are resisting big mining, oil, and timber interests in defense of their homes and communities. Poor neighborhoods in Chicago, New York, Los Angeles, Oakland, and around the country are coming together to build parks and community gardens and demand safe water and clean air for their children. African-American community leaders are not only expressing a deep commitment to environmental justice, they are reasserting their time-honored tie to the natural world. Communities that once saw environmentalism as a whites-only fight for national parks and clean beaches are now joining the struggle against environmental racism. And with the addition of each group, with the mobilization of each community, the movement grows richer and more powerful. The grassroots are seeding in diverse and sometimes unexpected places, and often with a clear rejection of the concession politics that has identified mainstream environmental organizations.

After a fair bit of criticism in the early 1990s, and some subsequent waffling, mainstream environmental organizations are starting to realize the need to greet the new face of environmentalism. Increasingly they're finding that the fight to protect wildlands, endangered species, water and air quality, can be complemented by adding to their repertoire fights for clean air in schools, worker safety, and greener and safer neighborhoods. They're beginning to appreciate more fully the connections between jobs, health, crime, education, poverty, and the environment. And, at a slightly slower pace, they're starting to recognize that these communities don't need to be given a program from on high; they understand the problems they face and they often have solutions, but they may be happy to have a helping hand.[17] Many environmental leaders are learning that it was a mistake to assume that African Americans, immigrant communities, and inner-city neighborhoods are not as concerned about the environment as middle-class whites.

In short, we're starting to discover an important truth – we're all in this together. This realization offers exciting possibilities for new solidarities and new articulations of environmental resistance. We can reject the tired divisions that cast neighborhood campaigns, the so-called 'not in my backyard' (NIMBY) fights, as somehow disconnected from the broader struggle for global change. The split between those who would fight against hunger or exploitation first and worry about our rivers and air later, and those who would switch those priorities no longer makes sense. More than this, as the environmental movement takes a new, global, multicultural, and multicolored face, it is likely to enrich our capacity to appreciate the link between exploitation and conservation. African Americans, who have borne the brunt of exploitation, know how power works and they can readily appreciate its global force. Immigrants who have come from impoverished countries are likely to see the links between the two worlds they have traveled. As all our lives and ideals grow increasingly degraded by corporate plunder, this new generation of environmentalists realizes that Bolivian peasants and inner-city mothers share a common vision. The lines that once separated poor and middle-class, urban and rural, black, brown, and white, are starting to blur.[18]

A quick rundown of a few of the most recent winners of the prestigious Goldman Environmental Prize (a sort of Nobel Prize for environmentalism) gives a sense of the character of the new global ecowarriors.[19] Margie Richard grew up literally next door to a Shell Chemical plant in a region of Louisiana known as Cancer Alley. After leading a thirteen-year community struggle and enduring countless abuses and corporate arrogance, she led the community to victory when Shell agreed to pay to relocate members of this underprivileged neighborhood to a safe area. Rashida Bee and Champa Devi Shukla of India led an international fight to hold Dow Chemical accountable for a major poison gas leak in Bhopal. Despite their poor health and poverty, they have ceaselessly fought for corporate accountability for a devastating act of corporate irresponsibility that killed 20,000 and left more than 150,000 seriously injured. In Colombia, in the wake of the environmental and social catastrophe brought on by years of violent conflict and exploitation, Afro-Colombian activist Libia Grueso won more than 5.9 million acres in territorial rights for the country's black rural peasantry. In Central Europe, Manana Kochladze worked to protect local villages and the countryside from an international consortium's plan to construct the world's largest pipeline through a national park and pristine mountains. After fighting fiercely for East Timor's liberation from Indonesian occupation, Demetrio do Amaral de Carvalho led the effort that won the inclusion of environmental justice tenets in his new country's constitution. This is the new global face of environmentalism.

These examples remind us that we need to re-imagine political action in a globalized world. Citizenship once meant responsible community participation at a local or perhaps national level; but this narrow view of political community is now wholly inadequate. Our everyday lives are entwined with the collective fate of the planet and peoples around the globe. We are citizens of the earth in every sense, and so we must redraw the borders of political accountability to match this reality. We've learned some valuable lessons over the past decades: it's now clear that saving our backyards alone is not possible; and the conflict between saving nature and saving people is a false one. We know that sustainability can only come with global equity. Saving rainforests in the Amazon, polar bears in the Arctic, women factory workers in India, and bushmen in the Kalahari are in fact the same struggle; and, in the end, they are all part of a struggle for our own survival. As the bumper stickers have it, we need to think globally and act locally; but we also need to think locally and act globally.

As the global crisis grows increasingly hard to ignore, the balance of forces around the world is shifting in favor of change. Poverty and insecurity are intensifying throughout the impoverished world and also among workers in wealthy countries. The environmental emergency is palpable in impoverished countries, where the devastation is felt every day through thirst, hunger, and disease. Increasingly, the crisis is also evident in wealthy countries, where fears of toxins in our food, air pollution warnings, contaminated shorelines, and the illness of our children are daily reminders of growing trouble. As a worldwide peoples' resistance takes form, the political stability of the present order comes ever more into question.

Talking about power

Changing the 'global system' is a tall order, and the forces that impose this system are entrenched and potent. But I'm optimistic. I'm optimistic because the existing order has a great weakness. Put plainly: the way we run our society is simply antithetical to our highest values. This hypocrisy, this deep moral inconsistency, is the toehold we need to demand change. It is only ignorance (intentional and otherwise) that allows the system to continue to appear legitimate and just. And it is time we attacked that ignorance. Policymakers, business leaders, and citizens can no longer be allowed to hide behind the claim that they were unaware or uninformed. We've seen a clear disjuncture between what Americans say matters and how they live their lives; we need to expose that dissonance. In short, we need to start talking about the things that the privileged and powerful would prefer we not discuss.

For example, let's consider the way our tax money is used to subsidize environmental destruction and poverty. Roughly $2.5 *trillion* in government funding every year goes to support the destruction of the global environment.[20] In the US, government subsidies for industrial agriculture, cheap water for manufacturing, logging on federal land, the fossil fuel industry, and automobiles amounts to $275 billion each year. We pay over $500 million every year to support the industrial ranching that destroys public rangelands. If government is supposed to work in our shared interests, you might expect it to support public transportation, small family farms, sustainable urban design and the like. But these get a tiny fraction of funding, if any. By one account, public subsidies for car culture amounts to about $2,500 per person each year.[21] At the same time, the federal government insists that the nation's railroad must pay for itself. It's bad enough that we don't adequately regulate pollution; we don't need to pay for it! It's time to get the automobile industry, coal power plants, and the mining and timber industries off corporate welfare. If we want to protect family farms (and a large majority of Americans say they do), let's enact policies that make it easier, not harder, for them to survive.

We can also talk about defining an economy as though the earth really matters. Any accountant will tell you that a bookkeeping method that counts all economic activity, no matter how destructive, in the plus column is ridiculous. We need to start measuring how well we're doing, not just how much we're consuming. And we need to adopt policies that maximize the former and minimize the latter. We can start by adopting policies that ensure that the price of items really represents the environmental and human cost of production. If we did this with gasoline, for example, internalizing the environmental and social costs as well as the massive public subsidies to the industry, a gallon of gas might cost as much as $15.[22] It's fairly clear how such a price shift would affect the momentum behind renewable energy and conservation. We don't need to wait for some technological breakthrough; we have the knowledge and resources to build a sustainable economy today.[23] All we need is the political will to make it happen.

It's time we considered defining policies that truly protect the things we hold dear. When companies violate the law and pump arsenic, mercury, and lead into our air, food, and water, we need to hit them hard. If public health matters to us, let's outlaw forms of production that are killing our children. For example, let's talk about the morality of a federal proposal to set up a trade in emission credits for the mercury pollution that is killing our children. The EPA has found that more than half a million newborns have unsafe levels of this neurotoxin in their bodies, yet in a proposal that was literally written by industry, the Bush administration suggested that we weaken restrictions and allow a trade in mercury emissions. I could go on with more examples, but my point is clear. It's time we exposed the gap between our deepest values and our public policy.

Let's talk about justice. It is simply not acceptable that the best predictor of where a hazardous waste site will be located is the surrounding community's race or ethnicity. It is deplorable that our children play in toxic playgrounds, try to learn in poisoned schools, and eat toxic foods. And it is an offense to anyone's basic sense of fairness that those who are the least able to protect themselves face the greatest risk. Hazardous waste is dumped on Native American reservations, poor farm workers are exposed to deadly poisons, and impoverished rural communities live in an increasingly lethal environment. It is time to demand justice.

Let's discuss the policies that seem to exist outside public perception, under a veil of ignorance that is carefully cultivated by corporate power. For example, the General Mining Act of 1872 allows corporations to purchase public lands for $5 an acre or less. Government managers don't have a choice. If the company wants to buy the land, the government must sell it at this cut rate. Companies don't even have to use it for mining; they just have to say they will. Western landscapes, sacred to native peoples, are left deeply scarred, with craters large enough to be seen from space, cyanide poisoning the groundwater, and millions of tonnes of waste littering the landscape. All this is subsidized by billions of our tax dollars. How many Americans know about this? Very few I would imagine. In California, industrial farmers get water from the federal government at about $3 for a thousand cubic meters (about 2 per cent of the market cost), even though it costs the government $25 to deliver it.[24] This virtually free water encourages farmers to grow water-intensive crops like alfalfa, rice, and cotton – crops that would otherwise in no way make sense in the arid California landscape. This hardly promotes sustainable water use! Would Americans like the fact that they are giving their money to the rich and powerful to promote such waste? I doubt it. These policies are outrageous. It's time Americans were outraged.

Demanding the unreasonable

It's hard to accept our inaction in the face of this growing global crisis. It is even harder to believe that as the crisis grows worse we will continue in our torpor. People often declare their conviction that eventually, before it's too late, humankind will take action to turn things around. 'People won't do anything until it gets really bad,' they'll declare. 'It's a shame,' they lament, 'but it takes a real urgent catastrophe to get people to act.' So they mix their lament with a blind optimism. Surely, at some point, before it's too late, we'll do *something*.

This reminds me of the boiling frog analogy. The story goes something like this. A teacher has two frogs. The first is gently lowered into a pan of water. The water is slowly heated; as the water gets gradually hotter, the frog adjusts a bit here and there and settles into a placid indolence. The water finally breaks into a boil and the frog dies. The second frog is tossed into a pot of already-boiling water; and it jumps out immediately. The lesson, of course, is that we are more easily seduced into acceptance when things move slowly. We adjust to new discomforts, accept them in small quantities and amend our expectations as necessary; but we are eventually overcome by the results of our own indolence. Well, I'm here to tell you that the water's boiling. Now is the time to act; and if we don't act now, there's no reason to hope that someone else will act in the future.

Perhaps you don't see the slow death of global climate change, toxic poisoning, hunger, and corporate exploitation as critical – hot water maybe, but not yet boiling. You may believe adjustment is still possible and even appropriate. But there are just too many examples of immediate and brutal devastation around the world to which no moral, decent person can adjust. The United Nations, the African Union, and activists around the world are begging the world to act with urgency to prevent an enormous crisis in Africa. James Morris, executive director of the United Nations' World Food Program, recently attempted to convince US senators to increase aid to the continent. 'AIDS and hunger interact,' he pointed out. 'AIDS dramatically undermines food production. Malnourished bodies are more receptive to HIV, and more receptive to the opportunistic diseases that follow.'[25] Over 30 million Africans are HIV-positive. Many more than this will die of hunger and associated illnesses if we continue to do nothing. Faced with rising public outrage, the finance ministers of the world's largest economies recently agreed to write off the debt of Africa's poorest countries. However, this offer came firmly attached to requirements for privatization and the removal of any impediments to foreign corporate endeavors in these countries. So, to save themselves from these most desperate conditions, they are made to embrace the same policies that have helped ensure corporate control and expand poverty and environmental catastrophe around the world for three decades. This is abhorrent.

Imagine for a moment trying to explain this ghastly situation to a visitor from another planet. Every day more than 5,000 Africans die of AIDS; another 11,000 become HIV-positive, a virtual death sentence in Africa. Another million or so Africans die of malaria each year. Yet another million children die from diarrhea every year. That's right – diarrhea! It is, of course, unthinkable that anyone should die of such a thing in the world's wealthy countries. But in African it is increasingly commonplace. And, of course, we have already talked about the plague of hunger. We have effective treatments for all these ills. Africa is dying. How would you explain this to a visitor who had no knowledge of the supposed economic 'realities' that legitimate our inaction? How will we explain this apparent indifference to future generations? What will we say when our children ask us what we did in the face of this immense calamity? How in the world could one possibly adjust to this horrible situation? The water is boiling.

Most Americans want their country to be a force for justice and peace in the world. And in a desire to be patriotic, too many simply believe that this is so, despite the evidence to the contrary. The effects of this cultivated public ignorance is evident, for example, in public perceptions of foreign aid. When questioned about foreign aid, most Americans initially believe the US spends too much. But, when asked to estimate how much the US spends, the average estimate in a recent survey was about 15 per cent of the federal budget. That's fifteen times the actual amount. When informed of the actual amount, the number who believe it is too much becomes a small minority.[26] The US in fact spends 0.1 per cent of GNP on foreign aid. That's less than 2 per cent of its spending on the military. It's also the least of any wealthy, industrialized country.

For the richest country in the world to be the stingiest in international aid is an abomination. Most of this aid isn't even for humanitarian purposes; it's politically driven. In fact, much of US foreign aid goes to relatively well-off countries; only about 20 per cent gets to the truly impoverished places that you probably imagine when you think about foreign aid. And even much of that aid stays in the hands of the rich in those countries. What's more, political pressure is mounting to cut this assistance significantly. Most disturbing of all, we continue to enforce a debt system that takes money out of the hands of the desperately poor to put in the hands of the ridiculously rich. This makes even the small amount of development aid that does exist virtually meaningless. The total foreign aid to the entirety of sub-Saharan Africa from all the world's wealthy countries combined amounts to only about one quarter of the region's total debt burden. Overall, wealthy countries take far more out of impoverished countries than they put into them. You're thinking this has got to change. But the reality is, in the current global order, it simply will not. We've got to change more than just the way we distribute foreign aid; we need to transform the fundamental values and priorities that define the way we do business.

You've read the book to this point, so you know all this. You know that the hunger–AIDS crisis is only one aspect of the global system of destruction and exploitation. The point I want to make now is that too many others don't know this. Or, when they are made aware, they shrug their shoulders or maybe shake their heads in pity, but somehow continue to believe that this is the natural state of things. Too many believe that this abominable condition is somehow inevitable, beyond human intervention. But massive hunger, slavery, sweatshops, and deadly thirst are not inevitable. Toxic waste is not an act of God. Exploitation is not a force of nature.

Some might claim that there's nothing we can do. They'll slip into cynicism, that incipient disease of the spirit. The truth is, we can change things. The good news about all the troubles we face is that we know how to fix them. We are able to end the global debt crisis whenever we decide that the lives of Africans are high enough on our list of priorities to do so. It's not physical or natural barriers that stop us from putting an end to hunger, only the dictates of the powerful and cultivated subjection of the many; both of these can be changed.

We know what needs to happen. We need to stop promoting and financing environmentally destructive mega-projects and start supporting projects that value and protect lives and ecosystems. We need to put sharp controls on corporate power and empower local peoples to define their own life options. We can start by defining a global living wage and outlawing the worst forms of labor abuse. We need to stop using poverty as leverage to expand corporate profits at the expense of human lives. We need to end a global debt system that undercuts democracy in favor of corporate control. We need to make preservation of the environment more profitable than its destruction. We need to move beyond the farcical international development aid now in place and begin promoting real sustainable development through well-funded and democratically accountable institutions.

Corporate leaders would like us to believe that such measures are unreasonable. Policies simply can't be enacted globally, they'll argue. We can't require an international living wage, they'll tell us. After all, they will plead, no one is really in charge of the global order. But this is simply not true. As you've seen, the international system is in fact tightly controlled; the WTO, the IMF, and the World Bank have the capacity to enforce their will around the globe. And these institutions are acutely accountable – accountable to corporate power brokers and their allies in public office. Unelected, unknown bureaucrats make binding decisions in opaque meetings in faraway places and the world is forever changed. People around the world will suffer the brunt of their decisions, but only the powerful are privy to the process. It is time that global policymakers are held accountable for their practices, not to corporations but to humanity.

It is a stark violation of the fundamental principles of sovereignty and democracy that the people that are most affected by the demands of institutions like the World Bank and the IMF have the least voice in defining their policies. Once again it is time that these institutions answer to the people upon whom they inflict their policies. It is time we democratize global governance. Some say that the IMF or WTO should be abolished. They are plainly unredeemable, they argue. Others call for reform of these organizations. Still others call for the formation of a World Environmental Organization, to counterbalance rather than eradicate these institutions and their influence. Whatever the case, it is clear that major change is needed and these changes can't be left up to corporate executives, bureaucrats, bankers, and free-market economists. We need an informed and inclusive public debate. It's high time that we, as a global society, decide what we want the future course of the world to look like.

The root of the problem is structural, but this doesn't mean that nothing short of global revolution will do the trick. Nobel prizewinning economist James Tobin has suggested that even a (tiny) 0.05 per cent tax on global speculative investments could not only make impoverished countries less vulnerable to global corporate forces, it could provide significant funds to fight hunger, disease, and environmental degradation.[27] Similarly, addressing global inequities in consumption will not mean dire poverty in the West. Simple and relatively inexpensive policies that build affordable and efficient mass transportation, discourage environmental waste, and increase investment in renewable energy could go a long way toward balancing the scales of global consumption. By one estimate, a full third of the US economy serves no worthwhile human purpose.[28] Globally, the total waste might exceed $10 trillion dollars. Ending this waste could free up a lot of resources to address the environmental and social crisis. One study concluded that if we shared the earth's available resources equitably, we could allow every person on the planet to live at about the level of the average, middle-class German.[29] Imagine an end to slavery, an end to hunger, and a major step toward a sustainable future for the planet. The cost? Forgoing the rabid consumption and work cycle that is driving most of us to despair, travelling on trains more often than cars, and spending more time with friends and family and less time shopping.

You say you want a revolution

The good news is that worldwide pressure for change is growing, and it is defining a global groundswell of popular resistance. Small farmers, textile workers, rural women, and others around the globe are coming together to resist.[30] Anti-globalization activists, anti-AIDS activists, peace activists, women, students, and workers have come together in the streets of New York, Tokyo, Buenos Aires, London, and Moscow to demand change. They hound global power brokers wherever they attempt to meet – Seattle, Washington DC, Quebec City, and Genoa. Each time tens if not hundreds of thousands take to the streets to voice their dissent. In response, leaders of the world's wealthy countries shift their meetings to remote Qatar, small villages in the Alps, or an isolated island off the coast of Georgia. The numbers that took to the streets to protest the Bush administration's war on Iraq were astounding. The message is clear: we are engaged, informed, and ready to fight for justice.

This isn't your father's protest movement. The current wave is different from historical protest movements in at least two ways. First, because of the enhanced coordination and communication made possible with the Internet, these activists are defining a truly global voice for change. In fact, the movement mirrors the Internet itself: linked affinity groups, multiple layers of organizational nodes, and a global web of communicative and coordinating interconnections, all radically decentralized but committed to a common vision. And, second, these activists have adopted a new vision of protests designed to inspire and enthuse the participants and the broader public. Protests often resemble celebrations. Whether it is an informal effort to 'Reclaim the Streets' from car culture with festive dance and music that merge civil disobedience with a carnival atmosphere, or the street theatre, costumes, floats, and puppets that have become standard fare at anti-WTO and peace protests, the mood is positive, celebratory, and filled with hope.

Critics accuse the protestors of oversimplifying complex problems, being too naive in their prescriptions, or having an unclear message. But the aim of protesters is never to lay out a detailed program. They use the power available to them, the capacity to attract public attention and maybe to disrupt business-as-usual for a few minutes, to jolt their concerns onto the public agenda. They agitate the complacent, provoke the tepid, and threaten the privileged. Most of all they confront the powerful with proof of their wrongs and confirmation of their hypocrisy.

As a result, protesters in the streets can shift the balance of the discussions in the conference rooms. The veteran activist and sociologist Todd

Gitlin outlined the indispensable mission of democratic crowds: 'to crack a wrongheaded consensus, to energize actual and potential reformers on the inside, to polarize opinion and goad laggards, to precipitate public debates that have been suppressed by establishments or pursued only by experts in closed rooms where inertia and groupthink overwhelm dissent.'[31] When public passions emerge, reform-minded policymakers on the inside gain a more attentive audience. The demands of the crowds become negotiating positions of the reformers. Some might call this co-optation. Maybe. But that's only a bad thing if the crowd outside stops pushing.

The breakdown of WTO talks in Cancún, Mexico, in 2003 demonstrated clearly the rising power of this de facto alliance. Protesters in the streets emboldened negotiators from impoverished counties to demand an end to a prime example of Western hypocrisy. Wealthy countries have insisted that poor nations eradicate protective tariffs and subsidies that support their agriculture and industry. As we've seen, using debt for leverage, they have often forced countries to do so. But simultaneously these same wealthy countries continue to subsidize their own agriculture to the tune of $300 billion annually. Taking a cue and encouragement from global protesters, representatives from Latin America, Asia, and Africa simply argued that what is good for the goose is good for the gander. They called for an end to massive European and American agricultural subsidies. (The US government spends about three times as much on cotton subsidies, for example, as it does on foreign aid to all of Africa.) These subsidies, they argued, deepen the hardship of farmers around the world by making them unable to compete with the subsidized industrial agriculture in wealthy countries; and, they added, such policies are quite simply blatantly unfair. Taken aback by this display of backbone, US and European negotiators left the meeting. They would have preferred not to talk about these duplicitous policies. But, like it or not, shouting in the street had now made subsidies a very real part of the discussion.

As we've seen, an important capacity of the powerful is to define the limits of public discussion and perception. Real power is not so much found in the ability to win an argument, but in stopping that argument from ever emerging. Public protests are a direct assault on this capacity. They aren't meant to win an argument, just to start one. So, after some 70,000 workers, environmentalists, and human rights advocates shut down the millennium round of WTO discussions in Seattle, President Clinton could no longer talk about free trade without also discussing labor rights and the environment. After 150,000 took to the streets in Genoa, world leaders had to start talking about global injustice. Similarly, mounting public criticism sparked by popular protests has forced the World Bank to re-evaluate the environmental consequences of their policies. Facing similar public condemnation, in 1999 the IMF committed itself for the first time to the aim of reducing global poverty. Of course, words are cheap and the battle is far from won; but shifting public attention and forcing issues onto the pubic agenda is a necessary first step.

Reasons for hope

What is taking shape may not be a revolution, but it may well foster revolutionary change. Innovative entrepreneurs are defining new possibilities, sustainable business models, and socially responsible practices. They are finding common cause with local governments and community groups determined to build sustainable alternatives to business as usual. These efforts are gaining impetus from community movements demanding environmental justice for their neighborhoods and sometimes for people half a world away that they will never meet. They join with the resistance of indigenous peoples, impoverished peasants, workers, slaves, women, and students around the globe. At the same time, protesters hound global power brokers, and insist that their voices be heard. And even within the ranks of the major environmental organizations there is an enthusiastic generation of new leaders, driven by a sense of justice and their deep community with nature, who are ready to push beyond the old politics of concession. These collective voices find resonance in the demands of human rights activists, feminists, workers' advocates, and many other contours of resistance. They are angry. They are willing to be impudent, unreasonable, and unpopular; and they are driven by a deep commitment to a just cause. They demand what they believe must be done, what is right and just, rather than what others say is practical or an appropriate compromise. This gives me great reason for hope.

Some argue that the lack of a distinct singular identity in this emerging resistance makes it unlikely to succeed. Without a doubt, the emerging resistance is defined by its wide array of priorities and broad range of approaches. But this is hardly a weakness. In fact, this may be its greatest strength. These agents of change are not driven by stale political dogma but by a need to defend their communities, their families, and the things that are precious to them. They are working alone and coming together to displace a corrupted system with hope. Green entrepreneurs redefining success in Germany, citizens groups demanding safe jobs in Chicago, concerned farmers mobilizing against a toxic waste dump in Mexico, indigenous peoples in South America refusing to give way to the bulldozers that would destroy their homes, worried mothers showing up at a school board meeting *en masse* to demand safe green spaces for their children, human rights activists in India marching against slavery, anti-corporate protesters in the streets of Moscow, London, Genoa and Seattle – all of them repeatedly express a reaffirmation of values they believe to be lost in contemporary society: compassion, balance, justice, and equity. And they all talk about democracy. They have no manifesto, but they do have an agenda. More than this, they have a shared vision that connects the politics of place, so central to the energized demands for justice emerging from communities around the world, with a political program for change that can be championed by established organizations, protesters in the streets, and reformers in the corridors of power. Environmentalism, properly understood, has the potential to crystallize a shared vision and unity of cause for the broader struggle for a just world. This is the movement's greatest challenge and greatest hope.

In the end, I believe we can do it. I believe we can re-forge the bonds of community that make us human, achieve justice, and enrich our lives in the process. I believe we can reject the vicious cycle of endless work for meaningless consumption. We can build a society that honors the earth, respects the balance of all life, and recognizes the precious dignity of all human beings. But I also believe that these changes will not come as the result of apolitical, individualized consumer choices. I don't believe shopping green will save us. And I don't believe these changes will come without struggle. It's time to demand justice. It's time to withdraw our consent. It's time we expose the hypocrisy, perversity, and suicidal nature of business-as-usual. This effort won't be spontaneous, it won't be clean, and it won't be romantic. We cannot afford to be naive. It will require negotiation, debate, and maybe a fair bit of screaming. But it is necessary, and we're running out of time. We've already lost too much. The race to the bottom must end.

Notes

chapter 1

1. *The FAO Measure of Chronic Undernourishment: What Is It Really Measuring?*, American Association for the Advancement of Science, Washington DC, 1997, p. 6.
2. UN, *The Cost of Poverty and Vulnerability*, Department of Economic and Social Affairs, United Nations, Geneva, 2001; World Bank, *Global Economic Prospects and the Developing Countries 2001*, World Bank, Washington DC, 2001, available at www.worldbank.org.
3. UNRISD, *States of Disarray: The Social Effects of Globalization*, United Nations Research Institute for Social Development, Geneva, 1997.
4. UNDP, *Human Development Report 2001*, United Nations Development Program, United Nations, Geneva, 2001, available at www.undp.org.
5. Ibid.
6. World Bank Statistical Information Management and Analysis (SIMA) database, available at www.worldbank.org.
7. UN, *The Cost of Poverty and Vulnerability*, p. 9.
8. UNDP, *Human Development Report 1998*, United Nations Development Program, United Nations, Geneva, 2001, available at www.undp.org.
9. UNDP, *Human Development Report 2001*.
10. SIMA database.
11. National Academy of Sciences, *Climate Change Science: An Analysis of Some Key Questions*, Committee on the Science of Climate Change, National Research Council of the United States, available at www.nap.edu.
12. US Environmental Protection Agency, *US Climate Action Report*, US Government Printing Office, Washington DC, 2002.
13. US Environmental Protection Agency, *Inventory of U.S. Greenhouse Gas Emissions and Sinks: 1990–1994*, US Government Printing Office, Washington DC, 1995, p. 3.
14. Ibid., p. 4.

15. Stephen H. Schneider, *Laboratory Earth: The Planetary Gamble We Can't Afford to Lose*, Basic Books, New York, 1997.
16. Committee on Abrupt Climate Change, Ocean Studies Board (OSB), Polar Research Board (PRB), *Abrupt Climate Change: Inevitable Surprises*, National Research Council, Washington DC, 2002, available at www.nap.edu.
17. Ibid.
18. Jay R. Malcom et al., *Habitats at Risk: Global Warming and Species Loss in Globally Significant Terrestrial Ecosystems*, World Wide Fund For Nature, Gland, 2002, p. 9; Susan Joy Hassol, *Arctic Climate Impact Assessment*, Cambridge University Press, Cambridge, 2004, p. 8.
19. James J. McCarthy and Malcolm McKenna, 'How Earth's Ice Is Changing,' *Environment*, vol. 42, no. 10 (December 2000): 8–19.
20. Myrna H.P. Hall and Daniel B. Fagre, 'Modeled Climate-Induced Glacier Change in Glacier National Park, 1850–2100,' *Bioscience*, vol. 53, no. 2 (February 2003): 131–41.
21. 'National Environmental Health Association Position on Global Climate Change,' *Journal of Environmental Health*, vol. 64, no. 2 (September 2001): 30–32.
22. J.T. Houghton et al., eds, *Climate Change 1995: The Science of Climate Change*, published for the Intergovernmental Panel on Climate Change by Cambridge University Press, New York, 1996; J.T. Houghton, Y. Ding, D.J. Griggs, M. Noguer, P.J. van der Linden, X. Dai, K. Maskell, and C.A. Johnson (eds), *Climate Change 2001: The Scientific Basis Contribution of Working Group I to the Third Assessment Report of the Intergovernmental Panel on Climate Change*, Cambridge University Press, Cambridge, 2001; John T. Houghton, *Global Warming: The Complete Briefing*, 2nd edn, Cambridge University Press, New York, 1997; J.T. Houghton et al., *Radiative Forcing of Climate Change*, Working Group I, Intergovernmental Panel on Climate Change, Cambridge University Press, New York, 1995. Or see United Nations Environment Programme and World Meteorological Organization, 'Why Do Human-made Greenhouse Gases Matter When Water Vapor is the Most Potent Greenhouse Gas?' in 'Common Questions about Climate Change,' available at www.gcrio.org/ipcc/.
23. Svein Tveitdal, 'Melting Permafrost May Accelerate Global Warming, UNEP Scientists Warn,' United Nations Environmental Program Press Release, February 2001.
24. 'Peat Bogs Harbour Carbon Time Bomb,' newscientist.com, 7 July 2004; C. Freeman et al., 'Export of Dissolved Organic Carbon from Peatlands under Elevated Carbon Dioxide Levels,' *Nature*, vol. 430, no. 6996 (2004): 195–8; Chris Freeman, Nick Ostle, and Hojeong Kang, 'An Enzymic "Latch" on a Global Carbon Store,' *Nature*, vol. 409, no. 6817 (2001): 149.
25. A.D. Friend, A.K. Stevens, R.G. Knox, and N.G.R. Cannell, 'A Process Based, Terrestrial Biosphere Model of Ecosystem Dynamics (HYBRID v 3.0),' *Ecological Modeling* 95 (1997): 249–87.
26. A.H. Johnson, A.J. Friedland, E.K. Miller, and T.G. Sicama, 'Acid Rain and Soils in the Adirondacks,' *Canadian Journal of Forest Research* 24 (1994): 663–9.
27. Intergovernmental Panel on Climate Change, IPCC IS 92a scenario, in J.T. Houghton et al. (eds), *Climate Change 1994: Radiative Forcing of Climate Change and an Evaluation of the IPCC IS 92 Emission Scenarios*, Cambridge University Press, Cambridge, 1995.
28. Richard P. Turco, *Earth Under Siege: From Air Pollution to Global Change*, Oxford University Press, Oxford, 1997.
29. Houghton, *Global Warming*, p. 110.
30. See www.epa.gov/globalwarming.
31. Ibid.
32. *Global Warming and Florida 2000–2100: A Summary of Reports from The Natural Resources*

Defense Council and The Union of Concerned Scientists – Ecological Society of America, Florida Climate Alliance, www.floridaclimatealliance.net.

33. Camille Parmesan and Gary Yohe, 'A Globally Coherent Fingerprint of Climate Change Impacts across Natural Systems,' *Nature*, vol. 421, no. 6918 (2003): 37; J. Barry, C. Baxter, R. Sagarin, and S. Gilman, 'Climate Related, Long-term Faunal Changes in a Californian Rocky Intertidal Community,' *Science*, vol. 267, no. 5198 (1995): 672–5; Gian-Reto Walther et al., 'Ecological Responses to Recent Climate Change,' *Nature*, vol. 416, no. 6879 (2002): 389; J.A. Pounds, M.P.L. Fogden, and J.H. Campbell, 'Biological Response to Climate Change on a Tropical Mountain,' *Nature*, vol. 398, no. 6728 (1999): 611–15.
34. Terry L. Root, Jeff T. Price, Kimberly R. Hall, Stephen H. Schneider, Cynthia Rosenzweig, and J. Alan Pounds, 'Fingerprints of Global Warming on Wild Animals and Plants,' *Nature*, vol. 421, no. 6918 (2003): 57–60; Camille Parmesan and Gary Yohe, 'A Globally Coherent Fingerprint of Climate Change Impacts across Natural Systems,' *Nature*, vol. 421, no. 6918 (2003): 37–42; A.T. Peterson et al., 'Future Projections for Mexican Faunas under Global Climate Change Scenarios,' *Nature*, vol. 416, no. 6881 (2002): 626–9; K. Reid and J.P. Croxall, 'Environmental Response of Upper Trophic-Level Predators Reveals a System Change in an Antarctic Marine Ecosystem,' Proceedings of the Royal Society of London B268 (2001): 377–84; V. Loeb et al., 'Effects of Sea-Ice Extent and Krill or Salp Dominance on the Antarctic Food Web,' *Nature*, vol. 387, no. 6636 (1997): 897–900.
35. T. Edward Mickens, 'North America's Fish Feel the Heat,' *National Wildlife*, vol. 40, no. 4 (June–July 2002): 42; see also www.epa.gov/globalwarming.
36. Phil Williamson, 'Disappearing Zooplankton,' *Planet Earth*, Autumn 2002; 'Atlantic Fish Crisis May Be Due to Global Warming,' *National Geographic News*, 9 January 2002, available at news.nationalgeographic.com/news.
37. See National Oceanic and Atmospheric Association, National Environmental Satellite, Data and Information Service 1998 and World Conservation Monitoring Centre, 1998. Also see International Society for Reef Studies (ISRS), numerous statements on global coral bleaching.
38. U. Irmler and V. Wiese, 'Ecological Impacts of Climate Change on National Parks and Protected Areas of the World,' World Wildlife Fund, Washington DC, 1997; J. Ellison, 'Mangrove Retreat with Rising Sea-level, Bermuda,' *Estuarine, Coastal, and Shelf Science* 37 (1993): 75–87.
39. 'Climate Claims the Golden Toad,' BBC, 26 April 1999, http://news.bbc.co.uk/1/hi/sci/tech/328776.stm; J. Alan Pounds, 'Climate and Amphibian Declines,' *Nature*, vol. 410, no. 6829 (2001): 639–40; J. Alan Pounds, Michael P.L. Fogden, and John H. Campbell, 'Biological Response to Climate Change on a Tropical Mountain,' *Nature*, vol. 398, no. 6728 (1999): 611–15.
40. Chris D. Thomas et al., 'Extinction Risk from Climate Change,' *Nature*, vol. 427, no. 6970 (2004): 145–8; J. Alan Pounds and Robert Puschendorf, 'Ecology: Clouded Futures,' *Nature*, vol. 427, no. 6970 (2004): 107–9; Peterson et al., 'Future Projections for Mexican Faunas': 626–9; A. Townsend Peterson, 'Projected Climate Change Effects on Rocky Mountain and Great Plains Birds: Generalities of Biodiversity Consequences,' *Global Change Biology*, vol. 9, no. 5 (2003): 647.
41. www.epa.gov/globalwarming.
42. World Meteorological Organization 2003, at www.wmo.ch.
43. Noam Mohr and Katherine Silverthorne, *Flirting with Disaster: Global Warming and the Rising Costs of Extreme Weather*, US Public Interest Research Group Education Fund, Washington DC, 1999.
44. 'Global Warming Said to Kill 160,000 a Year,' Cable News Network (CNN), 1 October 2003.

45. World Health Organization, 'Climate and Health,' www.who.int.
46. 'Cyclone Deaths Set to Pass 10,000,' BBC News online, 10 November 1999, available at news.bbc.co.uk.
47. United States Environmental Protection Agency, www.epa.gov/globalwarming.
48. *New York Times*, 13 December 2002.
49. C. Joly, 'Climate Change, Insurance and Investment Management,' in *Climate Change: Mobilizing Global Effort*, OECD, Paris, 1997, p. 53.
50. Ibid.
51. Aubrey Meyer, *Contradiction and Convergence: The Global Solution to Climate Change*, Green Books, London, on behalf of the Schumacher Society, 2000.
52. Houghton, *Global Warming*, p. 111.
53. Sari Kovats, Bettina Menne, Anthony McMichael, Roberto Bertollini, and Colin Soskolne (eds), *Climate Change and Stratospheric Ozone Effects on Our Health in Europe*, World Health Organization, Copenhagen, WHO Regional Publications, European Series, No. 88, p. 66, www.euro.who.int/.
54. Norman Myers, *Environmental Exodus: An Emergent Crisis in the Global Arena*, Climate Institute, Washington DC, 1995; Steve Lonergan, *Environmental Degradation and Population Displacement*, Global Environmental Change and Human Security Project, Research Report no. 1, Victoria BC, 1998.
55. Kovats et al., *Climate Change and Stratospheric Ozone Effects*, p. 66.
56. United Nations Food and Agricultural Organization, *The Sixth World Food Survey*, UN FAO, Rome, 1996, p. 16.
57. United Nations Environmental Program, *Global Environmental Outlook*, p. 242, at www.unep.org/geo/.
58. P.R. Shukla, 'Justice, Equity and Efficiency in Climate Change: A Developing Country Perspective,' in Ferenc Toth, ed, *Fair Weather?*, Earthscan, London, 1999.
59. International Energy Agency, *The Road from Kyoto: Current CO_2 and Transportation Policies in the IEA*, p. 120, at www.eceee.org/library_links/proceedings/2001.
60. Organization for Economic Cooperation and Development, 'Towards Sustainable Transportation,' conference organized by the OECD, hosted by the Government of Canada, 1996, p. 16, available at www.ecoplan.org.
61. OECD, *Motor Vehicle Pollution*, 1995, p. 37.
62. Ibid., p. 44.
63. Ibid., p. 89.
64. M. Santos, *The Environmental Crisis*, Greenwood, Westport CT, 1999, p. 72.
65. See World Meteorological Organization, *Scientific Assessment of Ozone Depletion: 2002*, UNEP/WMO, 31 July 2002.
66. Based on EPA report that 'A United Nations Environment Programme (UNEP) study shows that a sustained 1 percent decrease in stratospheric ozone will result in about a 2 percent increase in the incidence of non-melanoma skin cancer, which can be fatal,' and the claim that 'More than 1.2 million Americans will develop nonmelanoma skin cancer in 2000,' and that 'Currently, we are experiencing depletion [of ozone] of approximately 5 per cent at mid-latitudes.' J.D. Longstreth, F.R. de Guijl, M.L. Kripke, Y. Takizawa, and J.C. van der Leun, 'Effects of Increased Solar Ultraviolet Radiation on Human Health,' *Environmental Effects of Ozone Depletion: 1994 Assessment*, United Nations Environment Program, Nairobi, 1994.
67. J.F. Abarca and C.C. Casiccia, 'Skin Cancer and Ultraviolet-B Radiation under the Antarctic Ozone Hole: Southern Chile, 1987–2000,' *Photodermatology, Photoimmunology & Photomedicine*, vol. 18, no. 6 (2002): 294–303.
68. A. Blaustein, P. Hoffman, D.G. Hokit, J. Kiesecker, S. Wall, and J. Hayes, 'UV Retain and Resistance to Solar UV-B in Amphibian Eggs: A Link in Population Decline,' *Proceedings of the National Academy of Science* 91 (1994): 1791–5.

69. Dennis L. Hartmann, John M. Wallace, and Varavut Limpasuvan, 'Can Ozone Depletion and Global Warming Interact to Produce Rapid Climate Change?,' *Proceedings of the National Academy of Sciences of the United States of America*, vol. 97, no. 4 (2000): 1412–17.
70. D.T. Shindell, D. Rind, and P. Lonergan, 'Increased Polar Stratospheric Ozone Losses and Delayed Eventual Recovery Owing to Increasing Greenhouse-gas concentrations,' *Nature*, vol. 392, no. 6676 (1998): 589–92; 'Greenhouse Gases Contribute to Slower Ozone Layer Recovery,' *Environmental Science and Technology*, vol. 36, no. 9 (2002): 187; G. Pitari, E. Mancini, and V. Rizi, 'Impact of Future Climate and Emission Changes on Stratospheric Aerosols and Ozone,' *Journal of Atmospheric Sciences*, vol. 59, no. 3, pt. 1 (2002): 414–40; Scientific Assessment Panel of the Montreal Protocol on Substances that Deplete the Ozone Layer, *Scientific Assessment of Ozone Depletion: 2002*, UNEP/WMO, Nairobi, 31 July 2002.
71. Devra Davis, Alan Krupnick, and Geroge Thurston, 'The Ancillary Health Benefits and Cost of GHG Mitigation: Scope, Scale, and Credibility,' in D.L. Davis, A. Krupnick, and G. McGlynn (eds), *Proceedings of the Workshop on Estimating the Ancillary Benefits and Costs of Greenhouse Gas Mitigation Policies, March 27–29, 2000*, OECD, Washington DC, 2000, p. 137.
72. Luis Cifuentes, Victor H. Borja-Aburto, and Nelson Gouveia, 'Hidden Health Benefits of Greenhouse Gas Mitigation,' *Science*, vol. 293, no. 5533 (2001): 1257.
73. National Resources Defense Council, 'Top 50 MSA's Ranked by Attributable Mortality,' available at www.nrdc.org/air/pollution/bt/topMort.asp.
74. Davis, Krupnick, andMcGlynn (eds), *Proceedings of the Workshop on Estimating the Ancillary Benefits and Costs of Greenhouse Gas Mitigation Policies*, p. 145; Hillary French, 'You Are What You Breathe,' in Lester Brown, *The World Watch Reader on Global Environmental Issues*, W.W. Norton, New York, 1998.
75. 'E.P.A. Identifies 243 Counties That Fail Federal Air Standards,' Associated Press, 30 June 2004, p. 19.
76. Jane Vise Hall, 'Air Quality Policy in Developing Countries,' *Contemporary Economic Policy*, vol. 13, no. 2 (1995): 77.
77. Willine Carr and Lisa Zeitel, 'Variations in Asthma Hospitalizations and Deaths in New York City,' *American Journal of Public Health*, vol. 82, no. 1 (January 1992): 59–6; and W. Carr, L. Zeitel, and K. Weiss, 'Variations in Asthma Hospitalizations and Deaths in New York City,' *Pediatrics Supplement*, vol. 92, no. 2 (August 1993): 319–21.
78. K. Breslin, 'Focus: The Impact of Ozone,' *Environmental Health Perspectives*, vol. 103, no. 7–8 (1995): 660–64.
79. R.J. Earickson and I.H. Billick, 'The Geographic Variation in Urban Air Pollutants and Pediatric Blood Lead Levels,' *Applied Geography* 8 (1988): 5–23.
80. French, 'You Are What You Breathe'.
81. Ibid.
82. UNDP, *Human Development Report 2001*, Oxford University Press, Oxford, 2001, p. 1.
83. UNDP, *Human Development Report 1998*, p. 69.
84. Ibid.
85. R. Monastersky, 'Smoggy Asian Air Enters United States,' *Science News*, vol. 157, no. 1 (1 January 2000): 4.
86. R. Monastersky, 'The Stifling Side of Asian Exports,' *Science News*, vol. 155, no. 24 (12 June 1999): 383.
87. Tom Harner and Henrik Kylin, 'Polychlorinated Naphthalenes and Coplanar Polychlorinated Biphenyls in Arctic Air,' *Environmental Science & Technology*, vol. 32, no. 21 (November 1998): 3257–66.

88. Fred Pearce, 'Smog Controls Useless without Global Clean-Up,' *New Scientist*, vol. 176 (November 2002): 2365–78.
89. Janet Raloff, 'Global Smog: Newest Greenhouse Projection,' *Science News* 135, no. 17 (April 1989): 262–3; Heidi Ridgley, 'How Global Warming Affects Your Allergies,' *National Wildlife*, vol. 40, no. 3 (April–May 2002): 34.
90. For a review of acid pollution as a global phenomenon, see John McCormick, *Acid Earth: The Politics of Acid Pollution*, 3rd edn, Earthscan, London, 1997.
91. Chris Park, *Acid Rain: Rhetoric and Reality*, Methuen, New York, 1987; Inderjeet Sethi, M.S. Sethi, and S.A. Iqbal, *Environmental Pollution: Causes, Effects, and Control*, South Asia Books, Columbia MO, 1991; Marvin S. Soroos, *The Endangered Atmosphere: Preserving a Global Commons*, University of South Carolina Press, Columbia SC, 1997; D.C. Adriano and M. Havas, *Acidic Precipitation*, Springer Verlag, New York, 1990; B.J. Mason, *Acid Rain: Its Causes and Its Effects on Inland Waters*, Oxford University Press, New York, 1992.
92. E.D. Schulze and B. Ulrich, 'Acid Rain: A Large-Scale, Unwanted Experiment in Forest Ecosystems,' in Harold A. Mooney, Ernesto Medina, and David W. Schindler (eds), *Ecosystem Experiments*, John Wiley, New York, 1991, pp. 89–106; E.D. Schulze 'Air Pollution and Forest Decline in a Spruce Forest,' *Science*, vol. 224, no. 4906 (1989): 776–83.
93. See Bruce Forester, *The Acid Rain Debate*, Iowa State University Press, Ames, 1993; R.L. Arndt and G.R. Carmichael, 'Long-Range Transport and Deposition of Sulfur in Asia,' *Water, Air and Soil Pollution*, vol. 85, no. 4 (1995): 2283–8.
94. Martin Forstenzer, 'Ozone Damage to Forests: The Effects of Sprawl and Smog,' *Audubon* 96 (July–August 1994).
95. 'Pollution Said to Triple Easter Tree Deaths,' *New York Times*, 16 August 1994, c4; and Mary Mufford, 'Forest Talk on Coal River: Forest Crisis in Central Appalachia,' *American Forests* 101 (Autumn 1995): 16–17.
96. Gene Likens, Donald Buso, and Charles Driscoll, 'Long-Term Effects of Acid Rain: Response and Recovery of a Forest Ecosystem,' *Science*, vol. 272, no. 5359 (12 April 1996): 244–6; H.W. Zöttle and R.F. Hüttle (eds), *Management of Nutrition in Forests under Stress*, Kluwer Academic, Dordrecht, 1991.
97. Janet Raloff, 'When Nitrate Reigns: Air Pollution Can Damage Forests More than Trees Reveal,' *Science News* 147 (11 February 1995); A.H. Chappelka and P.H. Freer-Smith, 'Predisposition of Trees by Air Pollutants to Low Temperatures and Moisture Stress,' *Environmental Pollution*, vol. 87, no. 6 (1994): 105–17.
98. Richard P. Turco, *Earth under Siege: From Air Pollution to Global Change*, Oxford University Press, Oxford, 1997, p. 281.
99. Patricia Glick, 'The Toll from Coal,' *National Wildlife Federation*, April 2000, p. 1, available at www.nwf.org/washingtondc/resources.html.
100. C.T. Driscoll, G.B. Lawrence, A.J. Bulger, T.J. Butler, C.S. Cronan, C. Eagar, K.F. Lambert, G.E. Likens, J.L. Stoddard, and K.C. Weathers, *Acid Rain Revisited: Advances in Scientific Understanding since the Passage of the 1970 and 1990 Clean Air Act Amendments*, Hubbard Brook Research Foundation, Science Links Publication, vol. 1, no. 1, 2001.
101. J.L. Stoddard, D.S. Jeffries, and A. Lükewille, 'Regional Trends in Aquatic Recovery from Acidification in North America and Europe,' *Nature*, vol. 401, no. 6753 (1999), pp. 575–8; Janet Pelley, 'Belated Acid Rain Report May Trigger Legislation, Investigation,' *Environmental Science and Technology*, vol. 33, no. 11 (1999): 233–4.
102. Kevin Krajick, 'Long-Term Data Show Lingering Effects of Acid Rain,' *Science*, vol. 292, no. 5515 (13 April 2001): 195–6; G.E. Likens, C.J. Driscoll, and D.C. Buso, 'Longterm Effects of Acid Rain: Responses and Recovery of a Forest Ecosystem,' *Science*, vol. 272, no. 5259 (1996): 195.

103. S. McLaughlin and K. Percy, 'Forest Health in North America: Some Perspectives on Actual and Potential Roles of Climate and Air Pollution,' *Water, Air, and Soil Pollution*, vol. 116, no. 1–2, (November 1999): 151–97; G. Pitari, E. Mancini, V. Rizi, and D.T. Shindell, 'Impact of Future Climate and Emission Changes on Stratospheric Aerosols and Ozone,' *Journal of Atmospheric Science*, vol. 59, no. 3 (1 February 2002): 414–40; Eville Gorham, 'Lakes under a Three-Pronged Attack,' *Nature*, vol. 381, no. 6557 (1996): 109–10.
104. M. Posch, J.P. Hettelingh, J. Alcamo, and M. Krol, 'Integrating Scenarios of Acidification and Climate Change in Europe and Asia,' *Global Environmental Change*, vol. 6, no. 4 (1996): 375–94. J. Alcamo, M. Krol, and M. Posch, 'An Integrated Analysis of Sulphur Emissions, Acid Deposition, and Climate Change,' *Water, Air, and Soil Pollution* 85 (1996): 1539–50.
105. David W. Schindler, Curtis P. Jefferson, and Brian R. Parker, 'Consequences of Climate Warming and Lake Acidification for UV-B Penetration in North American Boreal Lakes,' *Nature*, vol. 379, no. 6567 (1996): 705–8.
106. Intergovernmental Panel on Climate Change, *Climate Change 1995 – The Science of Climate Change. Summary for Policymakers, and Technical Summary of the Working Group I Report*; IPCC, Geneva, 1996; Robert J. Charlson and Tom M.L. Wigley, 'Sulfate Aerosol and Climate Change,' *Scientific American* 270 (February 1994): 48–53.
107. J.P. Hetteling, M.J. Chadwick, H. Sverdrup, and D. Zhao, 'Assessment of Environmental Effects of Acidic Deposition in Asia,' in Wes Foell, Markus Amann, Greg Carmichel, Michael Chadwick, Jean-Paul Hettelingh, Leen Hordijik, and Zhao Dianwu (eds), *RAINS-ASIA: An Assessment Model for Air Pollution in Asia*, report on the World Bank-sponsored project, 'Acid Rain Emission Reduction in Asia,' December 1995; Wes Foell et al., 'Energy Use, Emissions and Air Pollution Reduction Strategies in Asia,' *Water, Air and Soil Pollution*, vol. 85, no. 4 (1995): 2277–82; H.E. Rodhe, E. Cowling, I.E. Galbally, J.N. Galloway, and R. Herrera, 'Acidification and Regional Air Pollution in the Tropics,' in H. Rodhe and R. Herrera (eds), *Acidification in Tropical Countries*, John Wiley, New York, 1988; Johan C.I. Kuylenstierna, Henning Rodhe, Steve Cinderby, and Kevin Hicks, 'Acidification in Developing Countries: Ecosystem Sensitivity and the Critical Load Approach on a Global Scale,' *AMBIO: A Journal of the Human Environment*, vol. 30, no. 1 (2000): 20–28.
108. Dhirendra K. Vajpeyi, 'Introduction,' in Dhirendra K. Vajpeyi (ed.), *Deforestation, Environment, and Sustainable Development: A Comparative Analysis*, Greenwood, New York, 2001, p. 3.
109. Emily Matthews, Richard Payne, Mark Rohweder, and Siobhan Murray, *Forest Ecosystems: Pilot Analysis of Global Ecosystems*, World Resources Institute, Washington DC, 2000; United Nations Food and Agriculture Organization, *Forest Resources Assessment 1990*, FAO, Rome, 1997, p. 3.
110. Food and Agriculture Organization, *Global Forest Resources Assessment*, FAO, Rome, 2000.
111. Derek Hall, 'Environmental Change, Protest, and Havens of Environmental Degradation: Evidence from Asia,' *Global Environmental Politics*, vol. 2, no. 2 (2002): 20–28.
112. Vajpeyi, *Deforestation, Environment, and Sustainable Development.*
113. United Nations Educational, Scientific and Cultural Organization, *Courier*, 10–15 September 1999.
114. Edward O. Wilson, *The Diversity of Life*, W.W. Norton, New York, 1992; Robert MacArthur and Edward O. Wilson, *The Theory of Island Biogeography*, Princeton University Press, Princeton NJ, 1967.
115. International Union for the Conservation of Nature and Natural Resources, *Red*

List of Threatened Animals, IUCN, Gland, available at www.redlist.org.

116. *Audubon Watchlist: An Early Warning System for Bird Conservation*, 2002, available at www.audubon.org/bird/watchlist/.
117. N.E. Stork, 'Measuring Global Biodiversity and its Decline,' in Marjorie L. Reaka-Kudla, Don E. Wilson, and Edward O. Wilson (eds), *Biodiversity II: Understanding and Protecting our Natural Resources*, National Academy of Science, Joseph Henry Press, Washington DC, 1997.
118. Wilson, *The Diversity of Life*, p. 7.
119. World Wildlife Fund (Worldwide Fund for Nature), *Living Planet Report 2002*, p. 2, available at www.panda.org.
120. David Spurgeon, 'Global Warming Threatens Extinction for Many Species,' *Nature*, vol. 407, no. 6801 (2000): 121.
121. Tropical Rain Forest Information Center, Michigan State University 'Rainforest Report Card,' available at www.bsrsi.msu.edu/rfrc/.
122. D.L. Skole, W.A. Salas, and C. Silapathong, 1998, 'Interannual Variation in the Terrestrial Carbon Cycle: Significance of Asian Tropical Forest Conversion to Imbalances in the Global Carbon Budget,' in J.N. Galloway and J.M. Melillo (eds), *Asian Change in the Context of Global Change*, Cambridge University Press, Cambridge, 1998, p. 163.
123. Matthews et al., *Forest Ecosystems*, p. 3.
124. T.E. Lovejoy, 'Biodiversity: What Is It?', in M.L. Reaka-Kudla, D.E. Wilson, and E.O. Wilson, (eds), *Biodiversity II: Understanding and Protecting Our Biological Resources*, Joseph Henry Press, Washington DC, 1997, pp. 2–4; and S. Schwartzman and M. Kingston, *Global Deforestation, Timber, and Struggle for Sustainability*, Environmental Defense, Washington DC, 1997.
125. Robert Morley, *Origin and Evolution of the Tropical Rain Forest*, Wiley, Chichester, 2000.
126. Shuzhen Sim and Noriko Toyoda, 'Rainforest Cures in Danger: The Malaysian Government Is Stealing the Traditional Knowledge of Indigenous People for Huge Profits,' Earth Island Institute, *The Borneo Project*, www.earthisland.org/borneo/news/wires/01win07.html.
127. Alan Durning, 'Cradles of Life,' in Brown, *The World Watch Reader on Global Environmental Issues*.
128. Cat Lazaroff, 'Rainforest Plants Help Battle Tuberculosis,' Environment News Service, 4 August 2000.
129. Brian Groombridge and Martin D. Jenkins, *World Atlas of Biodiversity: Earth's Living Resources in the 21st Century*, United Nations Environmental Program, World Conservation Monitoring Program, University of California Press, 2004.
130. Figures based on 1998–99, *World Resources: A Guide to the Global Environment*, World Resources Institute, UNEP, UNDP, and the World Bank, Washington DC, 1998.
131. Calculated from information at 'Rainforest Report Card,' www.bsrsi.msu.edu/rfrc/deforestation.html.
132. Carol Kaesuk Yoon, 'Something Missing in Fragile Cloud Forest: The Clouds,' *New York Times*, 20 November 2001, F4.
133. OECD, *Forestry, Agriculture and the Environment*, p. 124.
134. William Nester, *International Relations*, Wadsworth, Belmont CA, 2001.
135. Jeff McNeely and Sara Scherr, *Ecoagriculture: Strategies to Feed the World and Save Biodiversity*, Island Press, Washington DC, 2002.
136. 'Rainforest Report Card,' www.bsrsi.msu.edu/rfrc/deforestation.html.
137. Center for Disease Control, *Second National Report of Human Exposure to Environmental Chemicals*, www.cdc.gov.
138. Dana W. Kolpin, Edward T. Furlong, and Michael T. Meyer, 'Pharmaceuticals,

Hormones, and Other Organic Wastewater Contaminants in US Streams, 1999–2000: A National Reconnaissance,' *Environmental Science and Technology*, vol. 36, no. 6 (15 March 2002): 1202–11.

139. Environmental Protection Agency, *A Survey of Fish Contamination in Small Wadeable Streams in the Mid-Atlantic Region*, USEPA, National Exposure Research Laboratory, 2001, available at www.epa.gov/nerleerd/fishcontamsurv.pdf.
140. Karen Kidd, 'Effects of a Synthetic Estrogen on Aquatic Populations: A Whole Ecosystem Study,' *Toxic Substances Research Initiative*, Freshwater Institute, Winnipeg, 2003.
141. David Schindler at University of Alberta, as reported in *The Gazette* (Montreal) 6 October 1998.
142. 'Amazon Forest Growth Puzzles Scientists,' National Public Radio, *All Things Considered*, 10 March 2004.
143. Donella H. Meadows, Dennis L. Meadows, and Jørgen Randers, *Beyond the Limits: Confronting Global Collapse, Envisioning a Sustainable Future*, Chelsea Green, Post Mills VT, 1992.
144. Peter Meyer, Richard Williams, and Kristen Yount, *Contaminated Land: Reclamation, Redevelopment, and Reuse in the United States and the European Union*, Edward Elgar, Aldershot, 1995; Michael R. Edelstein, *Contaminated Communities: The Social and Psychological Impacts of Residential Toxic Exposure*, Westview Press, Boulder CO, 1988, p. 3.
145. Center for Environment, Health, and Justice, 'New Toxic Release Inventory (TRI) Data – Dioxin data reported for the first time. EPA's data shows that dioxin is being released into the environment at unsafe levels,' press release, 23 May 2002.
146. Peter Montague, 'Pesticides Pose Greater Threat to U.S. Drinking Water Supplies than Factories and Toxic Dumps,' *Rachel's Environment & Health News*, 1 December 1986.
147. Maude Barlow and Tony Clark, *Blue Gold: The Fight to Stop the Corporate Theft of the World's Water*, W.W. Norton, New York, 2003, pp. 28–9.
148. *Global Environmental Outlook Year Book 2003*, United Nations Environmental Program, Nairobi, 2004.
149. The Ocean Conservancy, www.oceanconservancy.org.
150. *Toxic Burden: PCBs in Marine Life*, Oceana, Washington DC, 2003, available at www.oceana.org/uploads/oceana_pop_final.pdf.
151. William L. Rathje and Cullen Murphy, *Rubbish! The Archaeology of Garbage*, University of Arizona Press, Tucson, 2001.
152. Debra L. Strong and Debi Kimball, *Recycling in America*, Santa Barbara CA, Abc-Clio, 1992.
153. Ibid.
154. Rathje and Murphy, *Rubbish!*, p. 206.
155. Carmela Federico, 'Dumping Computers Doesn't Compute!' www.envirolink.org.
156. Kirk Brown and K.C. Donnelly, 'An Estimation of the Risk Associated with the Organic Constituents of Hazardous and Municipal Waste Landfill Leachates,' *Hazardous Wastes and Hazardous Materials*, vol. 5, no. 1 (Spring 1988): 1–30.
157. United States General Accounting Office, *Siting of Hazardous Waste Landfills and Their Correlation with Racial and Economic Status of Surrounding Communities*, Government Printing Office, Washington DC, 1983.
158. United Church of Christ Commission for Racial Justice, *Toxic Wastes and Race in the United States: A National Report on the Racial and Socioeconomic Characteristics of Communities with Hazardous Wastes Sites*, Public Access, New York, 1987.
159. Mohai Paul and Bunyan Bryant, 'Environmental Racism: Reviewing the Evidence,' in Bunyan Bryant and Paul Mohai (eds), *Race and the Incidence of Environmental Hazards*,

Westview Press, Boulder CO, 1992: 163–76.
160. Marianne Lavelle and Marcia Coyle, 'Unequal Protection: The Racial Divide in Environmental Law,' *National Law Journal,* 21 September 1992.
161. R.J. Earickson and I. H. Billick, 'The Areal Association of Urban Air Pollutants and Residential Characteristics: Louisville and Detroit,' *Applied Geography* 8 (1988): 5–23; H.L. Needleman et al., 'Deficits in Psychological and Classroom Performance of Children with Elevated Dentine Lead Levels,' *New England Journal of Medicine* 300 (1979): 689–95.
162. Richard P. Turco, *Earth Under Siege: From Air Pollution to Global Change*, Oxford University Press, Oxford, 1997, p. 202.
163. John F. Lauerman, 'Wasting Away in the South Pacific,' *Environmental Health Perspectives*, vol. 109, no. 2 (February 2001), p. 66.
164. Valerie Brown, 'Old Pesticides Pose New Problems for Developing World,' *Environmental Health Perspectives*, vol. 109, no. 12 (December 2001), p. 578.
165. Human Rights Watch, 'Toxic Justice: Hunan Rights, Justice and Toxic Waste in Cambodia,' 1999, available at www.hrw.org.
166. Gary Cohen and John O'Connor (eds), *Fighting Toxics: A Manual for Protecting Your Family, Community, and Workplace*, National Toxics Campaign and Island Press, Washington DC, 1990, pp. 12–14.
167. Jonathan Loh (ed.), *Living Planet Report 2002*, WWF, Washington DC, 2002, p. 2.
168. UNDP, *Human Development Report 1998.*
169. Paul Ehrlich and Anne Ehrlich, *The Population Explosion*, Simon & Schuster, New York, 1990, p. 134.
170. Loh, *Living Planet Report 2002*, pp. 14–15.
171. Estimate based on consumption patterns in UNDP, *Human Development Report 2001* and *1998.*
172. UNDP, *Human Development Report 1998.*
173. Ibid.
174. P. Raskin, M. Chadwick, T. Jackson, and G. Leach, *The Sustainability Transition: Beyond Conventional Development*, Polestar Series Report no. 1, Stockholm Environmental Institute, 1996.
175. *World Development Report 1998*, World Bank, Washington DC, 1998.
176. Marilyn Beach, 'Water, Pollution, and Public Health in China,' *Lancet*, vol. 358, no. 9283 (2001): 735.
177. David G. Streets, Gregory R. Carmichael, and Markus Amann, 'Energy Consumption and Acid Deposition in Northeast Asia,' *Ambio*, vol. 28, no. 2 (1999): 135–43.
178. UNEP, *Global Environmental Outlook*, p. 224.
179. Richard Ellis, *The Empty Ocean: Plundering the World's Marine Life*, Shearwater Books, Washington DC, 2003.
180. UNEP, *Global Environmental Outlook*, p. 242.
181. Ibid.
182. Marq De Villiers, *Water: The Fate of Our Most Precious Resource*, Houghton Mifflin/ Mariner Books, New York, 2000.
183. UNEP, *Global Environmental Outlook*, p. 242.
184. Carmen Revenga, Jake Brunner, Norbert Henninger, Ken Kassem, Richard Payne, *Pilot Analysis of Global Ecosystems (PAGE): Freshwater Systems*, World Resource Institute, Washington DC, 2000, p. 27.
185. Vandana Shiva, *Water Wars: Privatization, Pollution, and Profit*, South End Press, Boston MA, 2003, p. 33.
186. *New York Times*, 24 November 2002.
187. UNEP, *Global Environmental Outlook.*
188. See Shiva, *Water Wars.*

189. 'Israel Warns of War over Water,' BBC News online, 10 September 2002.
190. Ismail Serageldin 'Of Water and Wars,' *Frontline* (India), vol. 16, no. 9 (24 April 1999), available at www.flonnet.com.
191. Michael T. Klare, *Resource Wars: The New Landscape of Global Conflict*, Owl Books, New York, 2002.
192. Peter Schwartz and Doug Randall, *An Abrupt Climate Change Scenario and Its Implications for U.S. National Security* (October 2003), report commissioned by the US Defense Department.

chapter 2

1. Clive Ponting, *A Green History of the World: The Environment and the Collapse of Great Civilizations*, Penguin, New York, 1993.
2. Francis Bacon, *The Great Instauration and New Atlantis*, Harlan Davidson, Arlington Heights IL, 1980 [1627], p. viii; also see Carolyn Merchant, *Death of Nature: Women, Ecology and the Scientific Revolution*, Harper, San Francisco, 1990, p. 169.
3. John Locke, *Second Treatise of Government*, Chapter V [1690], in John Locke, *On Politics and Education*, Walter Black, New York, 1947, p. 88.
4. Jeremy Rifkin, *Biosphere Politics: A Cultural Odyssey from the Middle Ages to the New Age*, Crown, New York, 1991; Gilbert Slater, *The English Peasantry and the Enclosure of the Common Fields*, Constable, London, 1907.
5. Marian Miller, 'Tragedy of the Commons,' in Dimitri Stevis and Valerie J. Assetto (eds), *The International Political Economy of the Environment: Critical Perspectives*, Lynne Rienner, Boulder CO and London, 2001.
6. Mike Davis, *Late Victorian Holocausts: El Niño Famines and the Making of the Third World*, Verso, London, 2001.
7. A.M. Mannion, *Global Environmental Change: A Natural and Cultural History*, Longman, London, 2nd edn, 1997, p. 118.
8. Emmerich de Vattel, *The Law of Nations: Book I: Of Nations Considered in Themselves*, ch. 1, 'Of Nations or Sovereign States', T. & J.W. Johnson, Philadelphia, 1883.
9. William Cronon, *Changes in the Land: Indians, Colonists and the Ecology of New England*, Hill & Wang, New York, 1983.
10. Henry Brackenridge, 'The Animals Vulgarly Called Indians,' *Freeman's Journal*, *c.* 1782; see also Henry Brackenridge, *Incidents of the Insurrection in the Western Parts of Pennsylvania, in the Year 1794*, College & University Press, New Haven CT, 1972.
11. Gary B. Nash, *Red, White, and Black: The Peoples of Early North America*, Prentice-Hall, Englewood Cliffs NJ, 1972, p. 136; also Donald A. Grinde, *Ecocide of Native America: Environmental Destruction of Indian Lands and Peoples*, Clear Light, Santa Fe NM, *c.* 1995.
12. Mannion, *Global Environmental Change*, p. 116.
13. In fact, residue of the increased pollutants from the mid-nineteenth century shift to coal can still be identified in Greenland ice caps. Samuel Hays, *A History of Environmental Politics since 1945*, University of Pittsburgh Press, Pittsburgh, 2000.
14. Ponting, *A Green History of the World*, p. 359; R.J. Flower et al., 'The Recent Acidification of a Large Scottish Loch Located Partly within a National Nature Reserve Site of Special Scientific Interest,' *Journal of Applied Ecology* 25 (1988): 715–24.
15. Acid rain was identified as early as the mid-nineteenth century by Robert Angus Smith, *Air and Rain: The Beginning of a Chemical Climatology*, Longmans Green, London, 1872. See R.J. Flower and R.W. Battarbee, 'Diatom Evidence for Recent Acidification of two Scottish Lochs,' *Nature*, vol. 305 (1983): 130–32; Flower et al., 'The Recent Acidification of a Large Scottish Loch'; R.W. Battarbee, 'Diatom

Analysis and the Acidification of Lakes,' *Philosophical Transactions of the Royal Society of London* 305 (1984): 451–77.
16. Anthony S. Travis, 'Contaminated Earth and Water: A Legacy of the Synthetic Dyestuffs Industry,' *Ambix*, vol. 49, no. 1 (2002): 21–50.
17. Peter Brimblecombe, *The Big Smoke: A History of Air Pollution in London since Medieval Times*, Methuen, London, 1987; A.S. Wohl, *Endangered Lives: Public Health in Victorian Britain*, Harvard University Press, Cambridge MA, 1983; Bill Luckin, *Pollution and Control: A Social History of the Thames in the 19th Century*, A. Hilger, Bristol, 1986; Lawrence Breeze, *The British Experience with River Pollution, 1865–1876*, P. Lang, New York, 1993; R. Hare, *Pomp and Pestilence: Infectious Disease, Its Origin and Conquest*, Philosophical Library, New York, 1954.
18. Stephan Boyden, *Western Civilization in Biological Perspective: Patterns in Biohistory*, Clarendon Press, Oxford, 1987.
19. Ponting, *A Green History of the World*, p. 368.
20. Friedrich Engels, *The Condition of the Working-Class in England in 1844*, Swan Sonnenschein, London, 1892, pp. 48–50.
21. William Cronon, *Nature's Metropolis: Chicago and the Great West*, W.W. Norton, New York, 1991.
22. Mannion, *Global Environmental Change*, p. 153.
23. Adam Markham, *A Brief History of Pollution*, St. Martin's Press, New York, 1994, p. 19.
24. Andrew C. Isenberg, *The Destruction of the Bison: An Environmental History, 1750–1920*, Cambridge University Press, New York, 2000.
25. George Perkins Marsh, *Man and Nature*, ed. David Lowenthal, Belknap Press, Cambridge MA, 1965, p. 36.
26. David M. Stiller, *Wounding the West: Montana Mining and the Environment*, University of Nebraska Press, Lincoln, 2000.
27. See, for example, W. Travis Hanes III and Frank Sanello, *Opium Wars: The Addiction of One Empire and the Corruption of Another*, Sourcebooks, Naperville IL, 2002.
28. John M. MacKenzie (ed.), *Imperialism and the Natural World*, Manchester University Press, Manchester, 1990; John Robert McNeill and Paul Kennedy, *Something New under the Sun: An Environmental History of the Twentieth-century World*, W.W. Norton, New York, 2000; Daniel J. Faber, *Environment under Fire: Imperialism and the Ecological Crisis in Central America*, Monthly Review Press, New York, 1993; Elinor G.K. Melville, *A Plague of Sheep: Environmental Consequences of the Conquest of Mexico*, Cambridge University Press, Cambridge, 1994; Tom Griffiths and Libby Robin (eds), *Ecology and Empire: Environmental History of Settler Societies*, University of Washington Press, Seattle, 1997; Madhav Gadgil and Ramachandra Guha, *This Fissured Land: An Ecological History of India*, University of California Press, Berkeley, 1993.
29. The legacy is clear today in Latin America and Africa. In Ecuador and Peru, 75 per cent of rural families do not have enough land to support themselves. In Egypt, a full 95 per cent are land-poor; in Latin America, two-thirds of the land is owned by less than 2 per cent of the population. Rehman Sobhan, *Agrarian Reform and Social Transformation*, Zed Books, London, 1993; Frances Moore Lappé, Joseph Collins, and Peter Rosset, *World Hunger: Twelve Myths*, 2nd edn, Grove Press for Food First, New York, 1998.
30. Adam Hochschild, *King Leopold's Ghost: A Story of Greed, Terror, and Heroism in Colonial Africa*, Houghton Mifflin, Boston MA, 1998.
31. Jan-Bart Gewald, *Herero Heroes: A Socio-political History of the Herero of Namibia, 1890–1923*, Ohio University Press, Athens, 1999; Peter H. Katjavivi, *A History of Resistance in Namibia*, Currey, London, 1988; Helmut Bley, *South-West Africa under German rule, 1894–1914*, trans. and ed. Hugh Ridley, Northwestern University Press,

Evanston IL, 1971.

32. Helge Kjekshus, *Ecology Control & Economic Development in East African History: The Case of Tanganyika, 1850–1950*, 2nd edn, Ohio University Press, Athens, 1996; L. Vail, 'Ecology and History: The Example of Eastern Zambia,' *Journal of Southern African Studies* 3 (1977): 129–55.
33. Piers Blaikie and Harold Brookfield with contributions by Bryant Allen et al., *Land Degradation and Society*, Methuen, New York, 1987.
34. Richard P. Tucker and J.F. Richards (eds), *Global Deforestation and the Nineteenth-century World Economy*, Duke University Press, Durham NC, 1983; David Arnold and Ramachandra Guha (eds), *Nature, Culture, Imperialism: Essays on the Environmental History of South Asia*, Oxford University Press, Delhi, 1995.
35. Richard P. Tucker, *Insatiable Appetite: The United States and the Ecological Degradation of the Tropical World*, University of California Press, Berkeley, 2000.
36. Paul Bairoch, 'International Industrialization Levels from 1750 to 1980,' *Journal of European Economic History* 11 (1982): 268–331.
37. J. Scott Parker, 'A Changing Landscape: Environmental Conditions and Consequences of the 1920s Union County Oil Booms,' *Arkansas Historical Quarterly*, vol. 60, no. 1 (2001): 31–52.
38. Lizabeth Cohen, *Making a New Deal: Industrial Workers in Chicago, 1919–1939*, Cambridge University Press, Cambridge, 1990.
39. Donald Worster, *Dust Bowl: The Southern Plains in the 1930s*, Oxford University Press, New York, 1979.
40. Murray Feshbach and Alfred Friendly, Jr, *Ecocide in the USSR: Health and Nature under Siege*, Basic Books, New York, 1992; see also Judith Shapiro, *Mao's War against Nature: Politics and the Environment in Revolutionary China*, Cambridge University Press, Cambridge, 2001.
41. Donald Worster, *Rivers of Empire: Water, Aridity, and the Growth of the American West*, Pantheon, New York, 1985.
42. John Muir, *The Yosemite*, Sierra Club Books, San Francisco, 1988 [1914].
43. Jeanne Nienaber Clark and Hanna J. Cortner, *The State and Nature: Voices Heard, Voices Unheard in America's Environmental Dialogue*, Prentice Hall, Upper Saddle River NJ, 2002, p. 178.
44. Organization for Economic Cooperation and Development, *Environmental Data Compendium 1999*, OECD, Paris, 1999, p. 230.
45. Jane Holtz Kay, *Asphalt Nation: How the Automobile Took over America and How We Can Take it Back*, University of California Press, Berkeley, 1997.
46. Ibid.
47. John Kenneth Galbraith, *Affluent Society*, Houghton Mifflin, Boston MA, 1958, p. 157.
48. Michael Redclift, *Wasted: Counting the Costs of Global Consumption*, Earthscan, London, 1996; Juliet Schor, *The Overworked American: The Unexpected Decline of Leisure*, Basic Books, New York, 1991; John de Graaf, David Wann, and Thomas H. Naylor, *Affluenza: The All-consuming Epidemic*, Berrett-Koehler, San Francisco, 2001.
49. UNDP, *Human Development Report 1998*.
50. Andrew Hurley, *Environmental Inequalities: Class, Race, and Industrial Pollution in Gary, Indiana, 1945–1980*, University of North Carolina Press, Chapel Hill, 1995.
51. Devra Lee Davis, *When Smoke Ran Like Water: Tales of Environmental Deception and the Battle against Pollution*, Basic Books, New York, 2002; William H. Helfand, Jan Lazarus, and Paul Theerman, 'Donora, Pennsylvania: An Environmental Disaster of the 20th Century,' *American Journal of Public Health*, vol. 91, no. 4 (2001): 553.
52. Chris Bryson, 'The Donora Fluoride Fog: A Secret History of America's Worst Air Pollution Disaster,' *Earth Island Journal*, vol. 13, no. 4 (1998): 36–8.

53. Davis, *When Smoke Ran Like Water.*
54. Markham, *A Brief History of Pollution*, p. 22.
55. Rachel Carson, *Silent Spring*, Houghton Mifflin, Boston MA, 1962, p. 13.
56. John Wargo, *Our Children's Toxic Legacy: How Science and Law Fail to Protect Us from Pesticides*, Yale University Press, New Haven, 1996; Barry Commoner, *The Closing Circle: Nature, Man and Technology*, Bantam Books, New York, 1972.
57. Feshbach and Friendly, *Ecocide in the USSR.*
58. Dick Thompson, 'The Greening of the U.S.S.R,' *Time*, 2 January 1989: 68
59. Douglas Stanglin, 'Toxic Wasteland,' *U.S. News & World Report*, 13 April 1992: 40.
60. Spencer R. Weart, 'Global Warming, Cold War, and the Evolution of Research Plans,' *Historical Studies in the Physical and Biological Sciences*, vol. 27, no. 2 (1997): 319–56.
61. National Cancer Institute, *Exposure of the American People to 131I from Nevada Atmospheric Bomb Tests*, NCI, Washington DC, 1997.
62. Albert O. Hirchman, *National Power and the Structure of Foreign Trade*, University of California Press, Berkeley, 1945.
63. On this point, of course, see Amartya Sen, *Poverty and Famines: An Essay on Entitlement and Deprivation*, Oxford University Press, Oxford, 1981.
64. Keith Griffin, *Alternative Strategies for Economic Development*, St. Martin's Press, New York, 1999.
65. Fernand Braudel, *Civilization and Capitalism*, vol. 3, Collins, London, 1981.
66. Malcom Waters, *Globalization*, New York: Routledge, 1995, p. 71.
67. UNDP, *Human Development Report 1999*, Oxford University Press for UNDP, New York, 1999.
68. World Bank, *World Development Indicators*, World Bank, Washington DC, 1998.
69. Thomas D. Lairson and David Skidmore, *International Political Economy: The Struggle for Power and Wealth*, 3rd edn, Thomson Wadsworth, Toronto, 2003, p. 78.
70. Ibid., p. 86.
71. Ibid., p. 90.
72. Robert K. Schaeffer, *Understanding Globalization: The Social Consequences of Political, Economic, and Environmental Change*, Rowman & Littlefield, Lanham MD, 1997, p. 65.
73. Anthony Payne, *Politics in Jamaica*, St. Martin's Press, New York, 1988.
74. On this point, see Ernst Friedrich Schumacher, *Small is Beautiful: A Study of Economics as if People Mattered*, Blond & Briggs, London, 1973.
75. Bill McKibben, 'A Special Moment in History,' *Atlantic Monthly*, May 1998: 55–78.
76. Jerry Mander, 'Intrinsic Negative Effects of Economic Globalization on the Environment,' in James Gustave Speth (ed.), *Worlds Apart: Globalization and the Environment*, Island Press, Washington DC, 2003.
77. Ibid., p. 118.
78. Schor, *The Overworked American.*
79. Ibid.
80. Juliet B. Schor, *The Overspent American: Why We Want What We Don't Need*, HarperCollins, New York, 1999.
81. De Graaf, Wann, and Naylor, *Affluenza*, p. 13.
82. Ibid., p. 2.

chapter 3

1. Joseph Stiglitz, *Globalization and Its Discontents*, W.W. Norton, New York, 2002, pp. 11–13.
2. Walden Bello, *Dark Victory: The United States and Global Poverty*, Pluto Press with

Food First and Transnational Institute, London, 1994, pp. 12–13.

3. United States Department of the Treasury, *The Multilateral Development Banks: Increasing U.S. Exports and Creating U.S. Jobs* (SuDoc T 1.2:B 22/10), Government Printing Office, Washington DC, 1994.
4. Nancy Alexander, Marcia Ishii-Eiteman, Ricardo Carrere, Aparna Sundar, Carol Welch, and Daphne Wysham, *Marketing the Earth: The World Bank and Sustainable Development*, Halifax Initiative and Friends of the Earth, 2002, available at www.halifaxinitiative.org or www.foe.org.
5. Stiglitz, *Globalization and Its Discontents*, p. 19.
6. Ibid., p. 73.
7. Jessica Woodroffe and Mark Ellis-Jones, *States of Unrest: Resistance to IMF Policies in Poor Countries*, World Development Movement Report, September 2000, available at www.globalpolicy.org.
8. Ibid.
9. Steven Morris, 'Unfair Trade Winds: What do Ecuadorean Bananas, Ugandan Coffee, and English Apples Have in Common? No Power,' *Guardian*, 17 May 2003.
10. American University Trade and Environment Database Case Studies, www.american.edu/TED/.
11. Timothy O'Brien and Margaret Kinnaird, 'Caffeine and Conservation,' *Science* 300 (2003): 587.
12. Timothy O'Brien and Margaret Kinnaird, of the Wildlife Conservation Society, as quoted in Steve Connor, 'Cheap Coffee Threatens to Wipe Out Wildlife and Ruin Farmers,' *Independent*, 26 April 2003.
13. Eswar Prasad, Kenneth Rogoff, Shang-Jin Wei, and M. Ayhan Kose, *The Effects of Financial Globalization on Developing Countries*, International Monetary Fund Occasional Paper, March 2003; Mohsin Khan, 'The Macroeconomic Effects of Fund-Supported Adjustment Programs,' *International Monetary Fund Staff Papers*, vol. 37, no. 2 (June 1990).
14. Eva Jespersen, 'External Shocks, Adjustment Policies and Economic and Social Performance,' in Giovanni Andrea Cornia, Rolph van der Hoeven, and Thandika Mkandawire (eds), *Africa's Recovery in the 1990s: From Stagnation and Adjustment to Human Development*, Macmillan, London, 1992; Robert Naiman and Neil Watkins, 'A Survey of the Impacts of IMF Structural Adjustment in Africa: Growth, Social Spending, and Debt Relief,' Center for Economic Policy Research, April 1999, available at www.cepr.net/publications/debt_1999_04.htm.
15. Alexander et al., *Marketing the Earth*, p. 10.
16. Naiman and Watkins, 'A Survey of the Impacts of IMF Structural Adjustment in Africa'.
17. 'The World Bank and the G7: Still Changing the Climate for Business,' Sustainable Energy and Economy Network (Institute for Policy Studies) and the International Trade Information Service, 1997–98, available at www.seen.org.
18. 'World Bank Working to Pave Way for Private Investment in China,' *Platts International Coal Report* 581 (30 September 2002): 8.
19. 'Social and Environmental Impact of Coal India Projects: IBRD Urged to Act Fast on Inspection Report,' *Business Line*, 12 February 2003.
20. 'Not Quite as Planned or Promised: The World Bank's Inspection Panel Finds that a Bank-backed Coal Mine Disrupted Livelihoods by Violating Guidelines on Resettlement, Supervision, and the Environment,' Bretton Woods Project, at www.brettonwoodsproject.org.
21. Current examples can be found at Amazon Watch, www.amazonwatch.org.
22. Alan Beattie, 'World Bank Defends Pipeline Support,' *New York Times*, 10 September 2002.

23. International Rivers Network, *Manibeli Declaration: Calling for a Moratorium on World Bank Funding of Large Dams*, September 1994, available at www.rivernet.org.
24. Michael M. Cernea, *The Risks and Reconstruction Model for Resettling Displaced Populations*, World Bank Environment Department, Washington DC, 1997; World Bank, *Resettlement and Development: The Bankwide Review of Projects Involving Involuntary Resettlement 1986–1993*, Environment Department Working Papers no. 32, Resettlement Series, 1994; World Bank, *Recent Experience with Involuntary Resettlement*, World Bank Operations Evaluation Study, World Development Sources, Washington DC, 1998–2000.
25. Michael M. Cernea, *African Involuntary Population Resettlement in a Global Context*, World Bank, Environment Department Working Papers no. 18174, Washington DC, 1997.
26. *Dams and Development: A New Framework for Decision-Making*, Report of the World Commission on Dams, 2000, available at www.dams.org.
27. According to the Argentine group Fundacion Vida Silvestre (Wildlife Foundation), see 'Argentine–Paraguay Dam May be Flooding Fragile Argentine Wetlands,' *EFE News Service*, 27 February 2003.
28. Chainarong Srettachau and Aviva Imhof, 'A Thai Dam, a Mistake, a Debt,' *Christian Science Monitor*, 9 August 2000.
29. See a series of reports by 'Mopheme,' 'the survivor,' including: 'Mafisa Releases Special Report On the Plight of Communities Affected By the LHWP', *Africa News*, 24 April 2003; '"LHDA Has Reneged on Its Promise of Rebuilding Our Houses Affected LHWP's Quarry," Says the HA NTSI Community,' *Africa News*, 13 March 2003.
30. In 1999, Health Care Without Harm (HCWH) released a report showing that at least thirty World Bank and International Finance Corporation (the private-sector arm of the World Bank) projects involved medical waste incineration. See *World Bank's Dangerous Medicine: Promoting Medical Waste Incineration in Third World Countries: An Inventory of World Bank Projects Involving Medical Waste Incineration*, Health Care Without Harm and Multinational Resource Center, June 1999, available at www.noharm.org; Neil Tangri, *Bankrolling Polluting Technology: The World Bank and Incineration*, Global Alliance for Incinerator Alternatives (GAIA), 2002, available at www.no-burn.org.
31. Bert Wilkinson, 'Canadian Mining Company Calls Class-action Suit "Unfounded" in Asking $2 Billion for 1995 Cyanide-tainted Spill,' *Associated Press*, 24 May 2003.
32. Greenpeace International, 'Sea Dumping of Wastes from the Mining Industry: The Case of the Lihir Gold Mine, Papua New Guinea,' Report for the London Convention, 24th Meeting, November 2002.
33. Mineral Policy Institute, *Lihir and International Law*, MPI Report 2002; and 'Australian Company's Toxic Waste Dumping Breaches International Law,' Mineral Policy Institute media release, 16 November 2002.
34. Program on Humanitarian Policy and Conflict Research, Harvard University, www.hsph.harvard.edu/hpcr/cpi/cpi.htm.
35. Chris McCall, 'Migration Fuels Indonesia's Fires, Fears,' *Reuters*, 19 December 1999.
36. World Bank, *The Forest Sector*, Policy Paper, 1991.
37. Marcus Colchester and Korinna Horta, *The World Bank and the World's Forests: A Lost Decade*, Forest Peoples Programme and Environmental Defense, 2000, available through the World Rainforest Movement at www.wrm.org.uy; World Rainforest Movement, 'Special Issue on the World Bank,' *WRM Bulletin*, 2000.
38. World Bank, *A Review of the World Bank's 1991 Forest Strategy and Its Implementation*, World Bank Operations Evaluation Department, Washington DC, 2000; and World Bank, 'The World Bank Forest Strategy: Striking the Right Balance', World Bank

Operation Evaluation Department, Washington DC, 2000, p. 23.

39. Jaroslava Colajacomo, 'Forests and the World Ban: No Need for a Step Back! Critical Forests, Protected Areas and Indigenous Rights at Risk in the New Draft of the World Bank Forest Policy,' Campagna per la riforma della Banca Mondiale (Campaign for the Reform of the World Bank), July 2002.
40. Jason Tockman, *The IMF: Funding Deforestation. How IMF Loans are Responsible for Global Forest Loss*, American Lands Alliance, 2001, available at www.americanlands.org.
41. Ibid.
42. Somewhere between $100 and $200 billion in investments each year is backed by these agencies. That's five to ten times the Bank's commitments. Aaron Goldzimer, 'Worse Than the World Bank? Export Credit Agencies – The Secret Engine of Globalization,' *Backgrounder Institute for Food and Development Policy*, vol. 9, no. 6 (Winter 2003).
43. Bruce Rich, 'Exporting Destruction,' *The Environmental Forum*, September 2000: 32–41.
44. Goldzimer, 'Worse Than the World Bank?'
45. 'Environment-China: Banned Voices Speak on Three Gorges Dam,' *Inter Press Service*, 12 December 2002.
46. John Pomfret, 'China's Monumental Gamble: River Halted by Dam That Will Bring Power, or Untold Destruction,' *Washington Post*, 2 June 2003.
47. International Rivers Network, Probe International, www.threegorgesprobe.org; Environmental Defense, www.environmentaldefense.org.
48. American University Trade and Environment Database, 'NAFTA and the Environment Case Analysis,' available at www.american.edu/TED/.
49. Ibid.
50. Hilary French, *Vanishing Borders: Protecting the Planet in the Age of Globalization*, W.W. Norton, New York, 2000.
51. David Bacon, 'NAFTA's Legacy – Profits and Poverty,' *San Francisco Chronicle*, 14 January 2004.
52. Sanford Lewis et al., *Border Trouble: Rivers in Peril, a Report on Water Pollution Due to Industrial Development in Northern Mexico*,: NTCF, Washington DC, May 1991; Texas Network for Environmental and Economic Justice, *Toxic Legacy: The Poisoning of West Dallas*, TNEEJ, Austin, 1994.
53. Gary Clyde Hufbauer and Jeffrey J. Schott, 'North American Environment under NAFTA,' *Institute for International Economics Paper*, October 2002; available at www.iie.com.
54. Robert Varady, Patricia R. Lankao, and Katherine Hankins, 'Managing Hazardous Materials Along the US–Mexico Border,' *Environment*, December 2001.
55. Cyrus Reed, Marisa Jacott, and Alejandro Vollamar, 'Hazardous Waste Management in the United States–Mexico Border States: More Questions than Answers,' *Red Mexicana de Accion Frente al Libre Cmercio*, March 2000; cited in Gary Clyde Hufbauer, Jeffrey J. Schott, Paul L.E. Grieco, and Yee Wong, *NAFTA Revisited: Achievements and Challenges*, Institute for International Economics, Washington DC, 2005, p. 28.
56. Janet Jarman, 'Cleaning Up Free Trade,' *Hemisphere: A Magazine of the Americas*, vol. 8, no. 2 (1998): 22–8.
57. Alberto Arroyo, 'Reflections from Mexico on the Myths of Free Trade,' in Sarah Anderson and Karen Hansen-Kuhn (eds), *Latin Americans against the FTAA: Another Americas is Possible*, Alliance for Responsible Trade, Washington DC, August 2001, available at www.art-us.org.
58. Celia Dugger, 'Report Finds Few Benefits for Mexico in NAFTA,' *New York Times*,

19 November 2003.
59. Howard Mann and Konrad von Moltke, *NAFTA's Chapter 11 and the Environment: Addressing the Impacts of the Investor–State Process on the Environment*, International Institute for Sustainable Development, Winnipeg, 1999.
60. Justin Gerdes, 'NAFTA's Chapter 11 threatens the Environment and Democracy,' *Environmental News Network*, 22 February 2002.
61. Alan Rugman, John Kirton, and Julie Soloway, *Environmental Regulation and Corporate Strategy: A NAFTA Perspective*, Oxford University Press, Oxford, 1999.
62. Marc Lee, 'Inside the Fortress: What's Going on at the FTAA Negotiations,' Canadian Centre for Policy Alternatives, April 2001.
63. Maude Barlow, *The Free Trade Area of the Americas and the Threat to Social Programs, Environmental Sustainability and Social Justice in Canada and the Americas*, Council of Canadians, 2002, available at www.canadians.org.
64. Martin Weber, 'Competing Political Visions: WTO Governance and Green Politics,' *Global Environmental Politics*, vol. 1, no. 3 (2001): 92–113.
65. Carmelo Ruiz, 'Green Protectionism,' *Earth Island Journal*, vol. 12, no. 2 (1997): 4–42.
66. Lori Wallach and Michelle Sforza, *The WTO: Five Years of Reasons to Resist Corporate Globalization*, Seven Stories Press, New York, 1999.
67. Steven Shrybman, *An Environment Guide to the World Trade Organization*, Canadian Alliance on Trade and Environment, Ottawa, May 1997.
68. National Research Council, *Decline of the Sea Turtles: Causes and Prevention*, National Academy Press, Washington DC, 1990.
69. Elizabeth R. DeSombre and J. Samuel Barkin, 'Turtles and Trade: The WTO's Acceptance of Environmental Trade Restrictions,' *Global Environmental Politics*, vol. 2, no. 1 (2002): 12–18.
70. Shalmali Guttal, 'Trading the Environment,' paper presented to seminar Alternatives to Neoliberalism, International South Group Network and the International Council of Social Welfare, 9 February 2000.
71. G. Kristin Rosendal, 'Impacts of Overlapping International Regimes: The Case of Biodiversity,' *Global Governance,* vol. 7, no. 1 (2001): 95–117; Graham Dutfield, 'Intellectual Property Rights, Trade and Biodiversity: The Case of Seeds and Plant Varieties,' International Union for the Conservation of Nature Background Paper, Intersessional Meeting on the Operations of the Convention, Montreal, 28–30 June 1999, available at www.iucn.org; J. Mugabe and E. Ouko, 'Control over Genetic Resources,' *Biotechnology and Development Monitor* 21 (1994): 6–7; Darrell A. Posey and Graham Dutfield, *Beyond Intellectual Property: Forward Traditional Resource Rights for Indigenous Peoples and Local Communities*, International Development Research Centre, London, 1996; Graham Dutfield, *Intellectual Property Rights, Trade and Biodiversity: Seeds and Plant Varieties*, Earthscan, Washington DC, 2002; Christophe Bellmann, Graham Dutfield, and Ricardo Melendez-Ortiz, *Trading in Knowledge: Development Perspectives on TRIPS, Trade and Sustainability*, Earthscan, London, 2004; and Biswajit Dhar and Sachin Chaturvedi, 'Implications of the Regime of Intellectual Property Protection for Biodiversity: A Developing Country Perspective' presented at Workshop on Biodiversity Conservation and Intellectual Property Regime, RIS/Kalpavriksh/IUCN, New Delhi, 29–31 January 1999.
72. *BRIDGES Weekly Trade News Digest*, vol. 2, no. 44 (16 November 1998).
73. Mike Williams, 'The Banana Trade: Subsidy Cut Peels Profits in Caribbean,' *Atlanta Journal and Constitution*, 5 August 2001.

chapter 4

1. Robert L. Nadeau, *The Wealth of Nature: How Mainstream Economics Has Failed the Environment*, Columbia University Press, New York, 2003; Brian Czech, *Shoveling Fuel for a Runaway Train*, University of California Press, Berkeley, 2000.
2. Gretchen C. Daily, *Nature's Services: Societal Dependence on Natural Ecosystems*, Island Press, Washington DC, 1997.
3. Eric A. Davidson, *You Can't Eat GNP: Economics as if Ecology Mattered*, Perseus Publishing, Cambridge MA, 2000.
4. Donella H. Meadows, *Beyond the Limits: Confronting Global Collapse, Envisioning a Sustainable Future*, Chelsea Green, Mills VT, 1992.
5. Jeremy Leggett, *The Carbon War: Global Warming and the End of the Oil Era*, Routledge, New York, 2001, p. 59.
6. Edward O. Wilson, *The Future of Life*, Vintage Books, New York, 2002, p. 24.
7. James Boyce, 'The Globalization of Market Failure?' in Kevin P. Gallagher and Jacob Werksman (eds), *The Earthscan Reader on International Trade and Sustainable Development*, Earthscan, London, 2002.
8. Ibid.
9. Paul Hawken, Amory Lovins, and L. Hunter Lovins, *Natural Capitalism: Creating the Next Industrial Revolution*, Little, Brown, Boston MA, 1999.
10. Environmental Justice Foundation, 'Oceans at Risk: The Real Cost of a Prawn Sandwich,' www.ecologyasia.com.
11. William J. Rea, 'Pesticides,' *Journal of Nutritional & Environmental Medicine*, vol. 6, no. 1 (1996): 55.
12. Frank Ackerman and Lisa Heinzerling, *Pricing the Priceless: On Knowing the Price of Everything and the Value of Nothing*, New Press, New York, 2004.
13. Mary O'Brien, *Making Better Environmental Decisions: An Alternative to Risk Assessment*, MIT Press, with Environmental Research Foundation, Cambridge MA, 2000.
14. Ibid.
15. Paul R. Hunter and Lorna Fewtrell, 'Acceptable Risk,' in Lorna Fewtrell and Jamie Bartram (eds), *Water Quality: Guidelines, Standards and Health*, World Health Organization, London, 2001.
16. Ackerman and Heinzerling, *Pricing the Priceless.*
17. David J. Ball, 'Environmental Risk Assessment and the Intrusion of Bias,' *Environment International*, vol. 28, no. 6 (2002): 529–44; and Ackerman and Heinzerling, *Pricing the Priceless.*
18. John Adams, 'Horse and Rabbit Stew,' in Annabel Coker and Cathy Richards, *Valuing the Environment: Economic Approaches to Environmental Evaluation*, Belhaven, New Haven, 1990.
19. Discount rate of 8 per cent compounded monthly.
20. Robert U. Ayres, 'How Economists Have Misjudged Global Warming,' *World Watch*, vol. 14, no. 5 (2001): 12–26.
21. Steven Weinberg, 'Mr. Bottom Line,' *OnEarth*, Spring 2003: 33–36; Ackerman and Heinzerling, *Pricing the Priceless.*
22. Article 10:6
23. C. Saladin, 'WTO "Supremacy Clause" in the POPs Convention,' Working Paper, Center for International Environmental Law, Washington DC, 1999.
24. Moreover, if the decrease does materialize, it's not at all clear at what level of development it takes place. Estimates of the per capita income where this downturn takes place range from $5,000 to $100,000. David I. Stern, 'Progress on the Environmental Kuznets Curve?' in Gallagher and Werksman, *International Trade and Sustainable Development.*

25. N. Shafik and S. Bandyopadhyay, 'Economic Growth and Environmental Quality: Time Series and Cross-country Evidence,' background paper prepared for World Bank, *World Development Report 1992*, Oxford University Press, New York, 1992; A. Antweiler, B. Copeland, and M.S. Taylor, *Is Free Trade Good for the Environment?*, NBER Working Paper 6707, National Bureau of Economic Research, Cambridge MA, 1998, available at www.nber.org/papers/.
26. Eyal Press, 'Jim Bob's Indonesian Misadventure,' *The Progressive*, June 1996: 34; see also Danny Kennedy with Pratap Chatterjee and Roger Moody, *Risky Businesss: The Grasberg Gold Mine. An Independent Annual Report on P.T. Freeport Indonesia, 1998*, Project Underground, 1998, available at www.moles.org/ProjectUnderground/.
27. Al Gedicks, *Resource Rebels: Native Challenges to Mining and Oil Corporations*, South End Press, Cambridge MA, 2001, p. 95.
28. Freeport–McMoran recently disclosed in a report to the Security Exchange Commission that it paid the Indonesian national military, Tentara Nasional Indonesia (TNI), an estimated US$5.6 million in 2002 for security purposes. 'Freeport–McMoran Admits Funding Millions to Indonesian Military,' *Drillbits and Tailings*, vol. 8, no. 3 (11 April 2003); see also Gedicks, *Resource Rebels*.
29. Raymond Bonner, 'U.S. Links Indonesian Troops to Deaths of 2 Americans,' *New York Times*, 30 January 2003.
30. Michael Conroy, Douglas Murray, and Peter Rosset, *A Cautionary Tale: Failed U.S. Development Policy in Central America*, Lynne Rienner, Boulder CO, 1996.
31. Arik Levinson and M. Scott Taylor, *Trade and the Environment: Unmasking the Pollution Haven Effect*, Working Paper, Center for Policy Research (CPR), Maxwell School of Syracuse University, February, 2001; Nick Mabey and Richard McNally, 'Foreign Direct Investment and the Environment: From Pollution Havens to Sustainable Development,' WWF–UK Research Paper, November 1999; Jennifer Clapp, 'What the Pollution Havens Debate Overlooks,' *Global Environmental Politics*, vol. 2, no. 2 (2002): 11–19.
32. Elliot Spagat, 'Mexican Power Plants Begin Operations amid U.S. Legal Fight,' *Associated Press*, 15 June 2003; Jennifer Clapp, *Toxic Exports: The Transfer of Hazardous Waste from Rich to Poor Countries*, Cornell University Press, Ithaca, 2001; Jennifer Clapp, 'Foreign Direct Investment in Hazardous Industries in Developing Countries: Rethinking the Debate,' *Environmental Politics*, vol. 7, no. 4 (1998): 92–113.
33. Barry Casteman, 'The Double Standards in Industrial Hazards,' in Jane H. Ives (ed.), *The Export of Hazard: Transnational Corporations and Environmental Control Issues*, Routledge & Kegan Paul, Boston MA, 1985.
34. Patrick Low and Alexander Yeats, 'Do "Dirty" Industries Migrate?' in Patrick Low (ed.), *International Trade and the Environment*, World Bank, Washington DC, 1992, pp. 89–103; Daniel W. Drezner, 'Bottom Feeders,' *Foreign Policy*, vol. 29, no. 6 (2000): 64–70; Joseph Kalt, 'The Impacts of Domestic Environmental Regulatory Policies on US International Competitiveness,' in A. Michael Spence and Heather A. Hazard (eds), *International Competitiveness*, Ballinger, Cambridge, 1988: 221–62.
35. Mabey and McNally, 'Foreign Direct Investment and the Environment.'
36. Muthukumara Mani and David Wheeler, 'In Search of Pollution Havens? Dirty Industry in the World Economy 1960–1995,' in Per G. Fredriksson (ed.), *Trade, Global Policy and the Environment*, World Bank, Washington DC, 1999: 115–27; Cees van Beers and Jeroen C J.M. van den Bergh, 'An Empirical Multi-Country Analysis of the Impact of Environmental Regulations on Foreign Trade Flows,' *Kyklas*, vol. 50, no. 1 (1997): 29–46.
37. Rolf Bommer, 'Environmental Policy and Industrial Competitiveness: The Pollution-Haven Hypothesis Reconsidered,' *Review of International Economics*, vol. 7, no. 2 (1999): 342–55.

38. 'A Job or Your Rights: Continued Sex Discrimination in Mexico's Maquiladora Sector,' *Human Rights Watch*, vol. 10, no. 1B (1998).
39. 'Maquiladoras Abuse Women's Rights,' *Human Rights Watch* (29 December 1998).
40. 'Mexico: The Shameful Side of the Maquiladoras,' International Confederation of Free Trade Unions, 2 November 1998.
41. Aaron Glantz, 'Jordan's Sweatshops: The Carrot or the Stick of US Policy?' *Corp-Watch*, 26 February 2003, available at www.corpwatch.org.
42. 'Labor Costs 11 Cents, The Sweatshirt $22.99,' *Daily News*, 18 April 2001, available at www.transnationale.org.
43. 'Labour Practices in Chinese Contract Factories Making Disney Products,' Hong Kong Christian Industrial Committee, February 2000, available at www.transnationale.org.
44. 'Secret Shame over Army of Child Workers,' *The Scotsman*, 15 July 2000.
45. *Fingers to the Bone: United States Failure to Protect Child Farmworkers*, Human Rights Watch, Washington DC, 2000, available at www.hrw.org.
46. 'Anti-union Repression in the Export Processing Zones,' International Confederation of Free Trade Unions, 1999, available at www.transnationale.org.
47. 'A Brief Profile of Free Trade Zones,' Employment Policies Coordinator, Centro de Investigacion para la Accion Femenina, Santo Domingo, Dominican Republic, 1996.
48. Kevin Bales, *Disposable People: New Slavery in the Global Economy*, University of California Press, Berkeley, 1999.
49. Human Rights Watch, 'Backgrounder: Child Labor in Agriculture,' *World Report 2002: Children's Rights*, at www.hrw.org.
50. Bales, *Disposable People*, ch. 4.
51. Ibid., pp. 3–4.
52. Kevin Bales and Jessica Reitz, 'Racism, Racial Discrimination and Related Intolerance Relating to Contemporary Forms of Slavery,' background paper prepared for OHCHR/UNESCO Workshop to Develop Publication to Combat Racism and to Foster Tolerance, Paris, 19–20 February 2003, United Nations High Commissioner for Human Rights, Paris, 2002, p. 4.
53. Bales, *Disposable People*.
54. Full text may be found at Global Situation Report, www.gsreport.com.
55. Laura A. Strohm, 'Pollution Havens and the Transfer of Environmental Risk,' *Global Environmental Politics*, vol. 2, no. 2 (2002): 29–36; David Wheeler 'Beyond Pollution Havens,' *Global Environmental Politics*, vol. 2, no. 2 (2002): 1–10; Jennifer Clapp, 'What the Pollution Havens Debate Overlooks,' *Global Environmental Politics*, vol. 2, no. 2 (2002): 11–19.

chapter 5

1. Michael Renner et al., *Vital Signs, 2003*, Worldwatch Institute, Washington DC, 2003.
2. United Nations Food and Agricultural Organization, 2002, www.fao.org.
3. Frances Moore Lappé, Joseph Collins, and Peter Rosset, *World Hunger: Twelve Myths*, 2nd edn, Grove Press for Food First, New York, 1998, p. 2.
4. Clive Ponting, *A Green History of the World: The Environment and the Collapse of Great Civilizations*, Penguin, New York, 1993, p. 254.
5. Lappé, Collins, and Rosset, *World Hunger*, p. 17.
6. Regression results for all countries outside the OECD, using IMF, World Bank, and FAO figures, show that there is a clear relationship between food exports per

capita and the percentage of the population who suffer malnutrition.

7. Juliet Gellatley with Tony Wardle, *The Silent Ark: A Chilling Expose of Meat – The Global Killer*, HarperCollins, New York, 1996, p. 181.
8. Results of a multivariate regression of foreign direct investment and gross domestic product with agricultural water use for non-OECD countries confirms this.
9. Maude Barlow and Tony Clark, *Blue Gold: The Fight to Stop the Corporate Theft of the World's Water*, W.W. Norton, New York, 2003, pp. 44–5; see also Sandra Postel, *Pillar of Sand: Can the Irrigation Miracle Last?* W.W. Norton, New York, 1999.
10. Regression of FDI and pesticide use in non-OECD countries, controlled for GDP.
11. S. Arpaia, 'Ecological Impact of Bt-transgenic Plants: Assessing Possible Effects of Cry IIIB Toxin on Honeybee (*Apis mellifera L.*) Colonies,' *Journal of Genetics and Breeding*, vol. 50, no. 4 (1996): 315–19; Nicholas Birch et al., 'Tri-trophic Interactions Involving Pest Aphids, Predatory 2–spot Ladybirds and Transgenic Potatoes Expressing Snowdrop Lectin for Aphid Resistance,' *Molecular Breeding*, vol. 5, no. 1 (1999): 75–83; Nizou Picard et al., 'Impact of Proteins Used in Plant Genetic Engineering: Toxicity and Behavioural Study in the Honeybee,' *Journal of Economic Entomology*, vol. 90, no. 6 (1997): 1710–16.
12. Susan Kegley, Lars Neumeister, and Timothy Martin, 'Disrupting the Balance Ecological Impacts of Pesticides in California,' Californians for Pesticide Reform and Pesticide Action Network North America, San Francisco, 1999.
13. David Pimentel and H. Acquay, 'Environmental and Economic Costs of Pesticide Use,' *Bioscience*, vol. 42, no. 10 (1992): 679–718.
14. E.G. Nielsen and L.K. Lee, 'The Magnitude and Costs of Groundwater Contamination from Agricultural Chemicals: A National Perspective,' Economic Research Service staff report, US Department of Agriculture, Natural Resource Economics Division, Washington DC, 1987.
15. Tracy Hewitt and Katherine R. Smith, *Intensive Agriculture and Environmental Quality: Examining the Newest Agricultural Myth*, Report from the Henry Wallace Institute for Alternative Agriculture, Greenbelt MD, 1995.
16. Howard Youth, *Winged Messengers: The Decline of Birds*, Worldwatch Paper 165, Worldwatch Institute, Washington DC, 2003.
17. Margaret Reeves, Kristin Schafer, Kate Hallward, and Anne Katten, 'Fields of Poison: California Farmworkers and Pesticides,' Pesticide Action Network North America, 2004, available at www.panna.org.
18. William J. Rea, 'Pesticides,' *Journal of Nutritional & Environmental Medicine*, vol. 6, no. 1 (1996): 55; S.D. Murphy, 'Toxic Effects of Pesticides,' in C.D. Klaassen, M.O. Amdur, and J. Doull (eds), *Toxicology: The Basic Science of Poisons*, 3rd edn, Macmillan, New York, 1986; C.F. Wilkinson and S.R. Baker (eds), *The Effect of Pesticides on Human Health*, Princeton Scientific Publications, Princeton NJ, 1990; World Health Organization/United Nations Environment Programme, *Public Health Impact of Pesticides Used in Agriculture*, WHO/UNEP, Geneva, 1989; P.T. Thomas and R.V. House, 'Pesticide-induced Modulation of Immune System,' in N.N. Ragsdale and R.E. Menzer (eds), *Carcinogenicity and Pesticides: Principles, Issues, and Relationships*, American Chemical Society, Washington DC, 1989, pp. 94–106; M.C. Fiore et al., 'Chronic Exposure to Aldicarb-contaminated Groundwater and Human Immune Function,' *Environmental Research* 41 (1986): 633–45; A.K. Saftlas et al., 'Cancer and Other Causes of Death among Wisconsin Farmers,' *American Journal of Industrial Medicine* 11 (1987): 119–29.
19. P. Rita, P.P. Reddy, and R.S. Venkatram, 'Monitoring of Workers Occupationally Exposed to Pesticides in Grape Gardens of Audhra Pradesh,' *Environmental Research* 44 (1987): 1–5; V.L. Georgieva, 'Cytogenetic Investigations in Agricultural Workers

in Occupational Contact with Pesticides,' in G. Szabo and Z. Papp (eds), *Medical Genetics: Proceedings of the Symposium at debrecen-Haldusƶobo-SƶLo, Hungary, 27–29 April 1976*, Exerpta Medica, Amsterdam, 1976, p. 297.

20. Bill Scanlon, 'Study Links Pesticides to Farmers' Depression,' *Mountain News*, 14 November 2002.
21. Food and Drug Administration, *Pesticide Program Residue Monitoring 2002*, Center for Food Safety and Applied Nutrition/Office of Plant and Dairy Foods, 2004, available at www.cfsan.fda.gov.
22. T. Holmes, E. Nielsen, and L. Lee., *Managing Groundwater Contamination in Rural Areas: Rural Development Perspectives*, US Department of Agriculture Economic Research Service, Washington DC, 1988.
23. National Water Quality Assessment Program at water.usgs.gov/nawqa/.
24. Robet J. Gilliom et al., 'Testing Water Quality for Pesticide Pollution,' *Environmental Science and Technology*, vol. 33, no. 7 (1999): 164–9; Rainer Fehr et al., 'Towards Health Impact Assessment of Drinking-water Privatization – The Example of Waterborne Carcinogens in North Rhine–Westphalia (Germany),' *Bulletin of the World Health Organiƶation*, vol. 81, no. 6 (2003): 408–16.
25. Irvin E. Liener, 'Toxins in Cow's Milk and Human Milk,' *Journal of Nutritional & Environmental Medicine*, vol. 12, no. 3 (2002): 175–88.
26. William J. Rea, 'Pesticides,' *Journal of Nutritional and Environmental Medicine*, vol. 6, no. 1 (1996): 55; David Pimentel and H. Acquay, 'Environmental and Economic Costs of Pesticide Use,' *Bioscience*, vol. 42, no. 10 (1992): 47–84.
27. D. Pimentel et al., 'Environmental and Economic Impacts of Reducing U.S. Agricultural Pesticide Use,' in D. Pimentel ed., *Handbook on Pest Management in Agriculture*, CRC Press, Boca Raton FL, 1991: 679–718.
28. Khabir Ahmad, 'United Nations Calls for Tighter Control on Pesticide Use in Poor Nations,' *Lancet*, vol. 360, no. 9345 (2002): 1574.
29. Angus Wright, 'Where does the Circle Begin? The Global Danger of Pesticide Plants,' *Global Pesticide Campaigner*, vol. 4, no. 4 (1994): 12.
30. Rea, 'Pesticides', p. 56.
31. WHO/UNEP, *Public Health Impact of Pesticides*; Ahmad, 'United Nations Calls for Tighter Control.'
32. Michael Eddleston et al., 'Pesticide Poisoning in the Developing World – A Minimum Pesticides List,' *Lancet*, vol. 360, no. 9340 (2002): 1163–8.
33. National Research Council, *Alternative Agriculture*, National Academy Press, Washington DC, 1989.
34. Pari Baumann, *Equity and Efficiency in Contract Farming Schemes: The Experience of Agricultural Tree Crops*, Working Paper 139, Overseas Development Institute, London, 2000; D. Glover and K. Kusterer, *Small Farmers, Big Business: Contract Farming and Rural Development*, St. Martin's Press, New York, 1990; P. Little and M. Watts (eds), *Living under Contract: Contract Farming and Agrarian Transformation in Sub-Saharan Africa*, University of Wisconsin Press, Madison, 1994; Sukhpal Singh, 'Theory and Practice of Contract Farming: A Review,' *Journal of Social and Economic Development*, vol. 3, no. 2 (2000): 228–46.
35. Roger Clapp, 'The Moral Economy of the Contract,' in Little and Watts (eds), *Living under Contract*, pp. 78–94; Michael J. Watts, 'Life under Contract: Contract Farming, Agrarian Restructuring, and Flexible Accumulation,' in ibid., pp. 21–77.
36. John Vidal, 'The Seeds of Wrath,' *Guardian*, 19 June 1999.
37. Alexandra M. Goho, 'Life Made to Order,' *Technology Review*, vol. 106, no. 3 (2003): 51–9.
38. Bill McKibben, *Enough: Staying Human in an Engineered Age*, Times Books, New York, 2003, back cover.

39. Daniel Charles and Brent Wilcox, *Lords of the Harvest: Biotech, Big Money, and the Future of Food*, Perseus Publishing, Cambridge MA, 2002.
40. Mark Schapiro, 'Sowing Disaster?' *The Nation*, 28 October 2002, available at www.thenation.com.
41. Bill Lambrecht, *Dinner at the New Gene Cafe: How Genetic Engineering is Changing What We Eat, How We Live, and the Global Politics of Food*, St. Martin's Press, New York, 2002.
42. Schapiro, 'Sowing Disaster?'
43. Stanley W.B. Ewen and Arpad Pusztai, 'Effect of Diets Containing Genetically Modified Potatoes Expressing Galanthus Nivalis Lectin on Rat Small Intestine,' *Lancet*, vol. 354, no. 9187 (1999): 1353–5.
44. Andrew Rowell, *Don't Worry [It's Safe to Eat]: The True Story of GM Food, BSE and Foot and Mouth*, London, Earthscan, 2003.
45. UNEP Global Biodiversity Assessment, www.unep-wcmc.org.
46. Andrew Kimbrell, *Fatal Harvest: The Tragedy of Industrial Agriculture*, Island Press, Washington DC, 2002.
47. FAO, *State of the World's Plant Genetic Resources for Food and Agriculture*, prepared for the 1996 Leipzig Conference; *Report of the FAO and Commission on Biodiversity*, Agricultural Biodiversity Workshop, Rome, 2–4 December 1998.
48. Robert F. Service, 'Seed-Sterilizing "Terminator Technology" Sows Discord,' *Science*, vol. 282, no. 5390 (1998).
49. Ibid., pp. 850–51.
50. Charles Benbrook, *Troubled Times amid Commercial Success for Roundup Ready Soybeans: Northwest Science and Environmental Policy Center: Glyphosate Efficacy is Slipping and Unstable Transgene Expression Erodes Plant Defenses and Yields*, Report Northwest Science and Environmental Policy Center, Sandpoint Idaho, 2001, available at www.biotech-info.net.
51. Rick Welsh, 'Testimony to the National Research Council's Board on Agriculture and Natural Resources Committee on Genetically Modified Pest Protected Plants,' Henry A. Wallace Institute for Alternative Agriculture, 24 May 1999, available at www.winrock.org.
52. Anne-Marie Chevre, F. Eber, A. Baranger, and M. Renard, 'Gene Flow from Transgenic Crops,' *Nature*, vol. 389 (1997): 924; P.J. Dale, 'Spread of Engineered Genes to Wild Relatives,' *Plant Physiology* 100 (1992): 13–15.
53. A.A. Snow, D. Pilson, L.H. Rieseberg, and M.J. Paulsen et al., 'A Bt Transgene Reduces Herbivory and Enhances Fecundity in Wild Sunflowers,' *Ecological Applications*, vol. 13, no. 2 (2003): 279–86; Ralph Haygood, Anthony Ives, and David Andow, 'Consequences of Recurrent Gene Flow from Crops to Wild Relatives,' *Proceedings of the Royal Society: Biological Sciences*, vol. 270, no. 1527 (2003): 1879–86; Norman C. Ellstrand, 'When Transgenes Wander, Should We Worry?' *Plant Physiology* 125 (2001): 1543–5; N.C. Ellstrand, H.C. Prentice, and J.F. Hancock, 'Gene Flow and Introgression from Domesticated Plants into their Wild Relatives,' *Annual Review of Ecological Systems* 30 (1999): 539–63.
54. G.R. Huxel, 'Rapid Displacement of Native Species by Invasive Species: Effect of Hybridization,' *Biological Conservation* 89 (1999): 143–52.
55. A.A. Snow and P. Palma, 'Commercialization of Transgenic Plants: Potential Ecological Risks,' *BioScience* 47 (1997): 86–96; R.S. Hails, 'Genetically Modified Plants: The Debate Continues,' *Trends Ecological Evolution* 15 (2000): 14–18.
56. M. Felke, N. Lorenz, and G.A. Langenbruch, 'Laboratory Studies on the Effects of Pollen from Bt-maize on Larvae of Some Butterfly Species,' *Journal of Applied Entomology*, vol. 126, no. 6 (2002): 20–25; R.L. Hellmich et al., 'Monarch Larvae Sensitivity to Bacillus Thuringiensis-purified Proteins and Pollen,' *Proceedings of the*

National Academy of Sciences 98 (2001): 11925–30; K.S. Oberhauser et al., 'Temporal and Spatial Overlap between Monarch Larvae and Corn Pollen,' *Proceedings of the National Academy of Sciences* 98 (2001): 11913–18; D.E. Stanley-Horn et al., 'Assessing the Impact of Cry1Ab-expressing Corn Pollen on Monarch Butterfly Larvae in Field Studies,' *Proceedings of the National Academy of Sciences* 98 (2001): 11931–6.

57. Margaret Mellon and Jane Rissler, 'Environmental Effects of Genetically Modified Food Crops,' *Food and Environment*, Union of Concerned Scientists, 2000, p. 5, available at www.ucsusa.org.
58. John Nichols, 'The Three Mile Island of Biotech?' *The Nation*, 30 December 2002.
59. Article in *Nature*.
60. David Ivanovich, 'Hope or Horror?' *Houston Chronicle*, 22 October 2000.
61. Michael Pollan, 'The Great Yellow Hype,' *New York Times*, 4 March 2001.
62. Tina Hesman, 'Bioengineered Rice Loses Glow as Vitamin A Source,' *St. Louis Post-Dispatch*, 4 March 2001.
63. Antony Barnett, 'US Accused of Piracy over "Mutant" Rice,' *Observer*, 28 October 2001.
64. John Madeley, 'Commodities and Agriculture: Plant Patents Raise Fear of "Biopiracy": Rights to Basmati Rice and the Potato Have Come into Question,' *Financial Times*, 8 December 2000.
65. Andrew T. Mushita and Carol B. Thompson, 'Patenting Biodiversity? Rejecting WTO/TRIPS in Southern Africa,' *Global Environmental Politics*, vol. 2, no. 1 (2002): 65–82.
66. William Allen, 'Lant-Derived Medicines Raise Issue of Ownership,' *St. Louis Post–Dispatch*, 22 December 1991.
67. Action Group on Erosion, Technology and Concentration (ETC), 'Peruvian Farmers and Indigenous People Denounce Maca Patents: Extract of Andean Root Crop Patented for Natural Viagra Properties,' www.etcgroup.org.
68. Dina Kraft, 'Bushmen, Drug Giant Battle over Royalties,' *Toronto Star*, 24 August 2002.
69. Mushita and Thompson, 'Patenting Biodiversity?'; R.J. McNeil and M.J. McNeil, 'Ownership of Traditional Medicines: Moral and Legal Obligations to Compensate for Taking,' *North-East Indian Quarterly*, Autumn 1989; M.J. Huft, 'Indigenous People and Drug Discovery Research: A Question of Intellectual Property Rights,' *North Western University Law Review* 89 (1995).
70. *World Water Development Report: Water for People, Water for Life*, UNESCO Division des Sciences de l'Eau, Paris, 2003, available at www.unesco.org/water/wwap/wwdr, calculated from data provided in ch. 1.
71. UNEP, *Groundwater and its Susceptibility to Degradation: A Global Assessment of the Problem and Options for Management*, available at www.unep.org/DEWA/water/groundwater.
72. World Meteorological Organization (WMO), *Comprehensive Assessment of the Freshwater Resources of the World*, WMO, Geneva, 1997, p. 9.
73. Central Intelligence Agency, *Global Trends 2015: A Dialogue About the Future with Nongovernment Experts*, Government Printing Office, Washington DC, 2000.
74. *Water for People, Water for Life*, p. 18; see also WMO, *Comprehensive Assessment of the Freshwater Resources of the World.*
75. 'Water Incorporated: The Commodification of the World's Water,' *Earth Island Journal*, 5 March 2002.
76. World Commission on Dams, *Dams and Development: A New Framework for Decision-Making*, WCD Report, 16 November 2000.
77. WMO, *Comprehensive Assessment of the Freshwater Resources of the World*, p. 9.

78. Mark W. Rosegrant, *Water Resources in the Twenty-First Century: Challenges and Implications for Action, Food, Agriculture, and the Environment*, Discussion Paper 20, International Food Policy Research Institute, Washington DC, 1997, p. 4.
79. Lawrence J. MacDonnell, *From Reclamation to Sustainability: Water, Agriculture, and the Environment in the American West*, University Press of Colorado, Denver, 1999.
80. Ibid.
81. Aral Sea website, www.dtd.dlr.de/app/land/aralsee/.
82. George W. Kling et al., *Confronting Climate Change in the Great Lakes Region: Impacts on our Communities and Ecosystems*, Report from the Union of Concerned Scientists, Washington DC, 2003.
83. United Nations Environment Programme (UNEP), Division of Environment Information and Assessment, *Characterization and Assessment of Groundwater Quality Concerns in Asia-Pacific Region*, UNEP, Nairobi, 1996.
84. World Health Organization (WHO), *Health and Environment in Sustainable Development: Five Years After the Earth Summit*, WHO, Geneva, 1997.
85. United States Geological Survey, *The Quality of Our Nation's Waters: Nutrients and Pesticides*, Government Printing Office, Washington DC, 1999; Bernard T. Nolan and Jeffrey D. Stoner, 'Nutrients in Groundwaters of the Coterminous United States, 1992–1995,' *Environmental Science and Technology*, vol. 34, no. 7 (2000): 1156–65.
86. Payal Sampat, 'Uncovering Groundwater Pollution,' in Lester Brown et al. (eds), *State of the World 2001*, W.W. Norton for the Worldwatch Institute, New York, 2001, p. 22.
87. Quoted in 'Water Incorporated'.
88. Tony Clarke and Maude Barlow, 'Blue Gold Rage Challenging the Corporate Takeover of Water Systems in Latin America,' Water Rights Project, Polaris Institute, Ottawa, 2003, available at www.polarisinstitute.org; see also Maude Barlow and Tony Clarke, *Blue Gold: The Global Water Crisis and the Commodification of the World's Water Supply,* New York: New Press, 2001.
89. As cited in Vandana Shiva (Director of the Research Foundation for Science, Technology and Ecology, New Delhi), 'Monsanto's Expanding Monopolies,' *The Hindu*, 1 May 1999.
90. Ibid.
91. Shawn Tully, 'Water, Water Everywhere,' *Fortune*, 15 May 2000.
92. Public Service International Research Unit Report, available at www.psiru.org; Kate Bayliss, 'Water Privatization in SSA: Progress, Problems and Policy Implications,' paper presented at the Development Studies Association Annual Conference, University of Greenwich, 9 November 2002.
93. Bob Carty, 'Whose Hand on the Tap? Water Privatization in South Africa,' CBC Radio, February 2003, www.cbc.ca; Maude Barlow and Tony Clarke, 'Water Apartheid: Privatization in South Africa,' *The Nation*, September 2002.
94. 'Water Incorporated.'
95. Barlow and Clarke, *Blue Gold.*
96. Ibid.
97. Sandra Postel, *Last Oasis: Facing Water Scarcity*, Worldwatch Institute, London, 1992.

chapter 6

1. George A. Gonzales, *Corporate Power and the Environment: The Political Economy of US Environmental Policy*, Rowman & Littlefield, Lanham MD, 2001.
2. J. Brooks Flippen, *Nixon and the Environment*, University of New Mexico Press,

Albuquerque, 2000.

3. Gonzales, *Corporate Power and the Environment.*
4. James Miller, 'Air Pollution,' in *Nixon and the Environment: The Politics of Devastation*, Taurus Communications, New York, 1972.
5. Gonzales, *Corporate Power and the Environment*.
6. James Watt, 5 February 1981; cited in Lonnie Murray, 'Spirituality and Wetland Conservation,' *Nature Spirit*, 30 April 2004.
7. *The Economist*, 5 March 1983, p. 39.
8. Matthew Cahn, *Environmental Deceptions: The Tension between Liberalism and Environmental Policymaking in the United States*, State University of New York Press, Albany, 1995.
9. Ibid., p 61.
10. Sheila Cavanagh, Robert Hahn, and Robert Stavins, 'National Environmental Policy During the Clinton Years,' Faculty Research Working Papers, John F. Kennedy School of Government Working Paper, 2001.
11. *Washington Post*, 22 April 1993.
12. Jeffrey St. Clair, *Been Brown So Long, It Looked Like Green to Me: The Politics of Nature*, Common Courage Press, Monroe ME, 2003.
13. Guy Gugliotta and Eric Pianin, 'EPA: Few Fined for Polluting Water,' *Washington Post*, 6 June 2003.
14. Office of Inspector General, Environmental Protection Agency (EPA), *Water Enforcement: State Enforcement of Clean Water Act Dischargers Can Be More Effective*, Report No. 2001–P-00013, August 2001, p. ii.
15. Ibid., p. iv.
16. National Water Quality Inventory: 1998, Report to Congress, available at www.epa.gov.
17. Victor B. Flatt, 'A Dirty River Runs Through It (The Failure of the Enforcement of the Clean Water Act),' *Boston College Environmental Affairs Law Review*, vol. 25, no. 1 (1997): 1–45.
18. EPA, *Water Enforcement*, p. 17.
19. Council on Environmental Quality, 'Environmental Quality–1980,' US Government Printing Office, Washington DC, 1980.
20. Gonzales, *Corporate Power and the Environment.*
21. Ibid.
22. Vanessa Houlder, 'Fighting To Be Heard: In Spite of their Real Power, Pressure Groups Feel Increasingly Beleaguered,' *Financial Times*, 20 September 2000.
23. Nancy Caron and Don Munton, 'Flaws in the Conventional Wisdom on Acid Deposition,' *Environment*, vol. 42, no. 2 (2000): 33–5.
24. Still, the Clean Air Act wasn't entirely irrelevant. It allowed plants to renegotiate their contracts with coal companies using the new federal requirements as a legal excuse. A few utilities offset some of the cost of scrubbers by selling excess emission allocations.
25. *Acid Rain Revisited: Advances in Scientific Understanding since the Passage of the 1970 and 1990 Clean Air Act Amendments*, Hubbard Brook Research Foundation, Hanover NH, 2001, available at www.hbrook.sr.unh.edu.
26. *Asleep at the Wheel: The Environmental Protection Agency's Failure to Enforce Pollution Standards for Heavy-Duty Diesel Trucks*, Staff Report, Committee on Commerce US House of Representatives, March 2000.
27. John DeGaspari, 'Retooling CAFÉ,' *Mechanical Engineering*, April 2004.
28. Mark Dowie, *Losing Ground: American Environmentalism at the Close of the Twentieth Century*, MIT Press, Cambridge MA, 1996.
29. Ibid.

30. John Bellamy Foster, *Ecology against Capitalism*, Monthly Review Press, New York, 2002.
31. Maria Weigner and Nancy Watzman, *Paybacks: Policy, Patrons, and Personnel*, Joint report by Earth Justice and Public Campaign, September 2002.
32. *Wall Street Journal*, 13 June 2001, p. 9, cited in ibid.
33. US Environmental Protection Agency, *Unfinished Business: A Comparative Assessment of Environmental Problems*, EPA, Washington DC, 1987.
34. *Development and Environment*, UN General Assembly Resolution, A/RES/2849 (XXVI), 17 January 1972.
35. Lynton Caldwell, *International Environmental Policy: Emergence and Dimensions*, Duke University Press, Durham NC, 1984.
36. UN, *United Nations' World Commission on Environment and Development*, Oxford University Press, Oxford, 1987.
37. United Nations, *World Development Report, 1992*, p. 4.
38. Paul Harris, *International Equity and Global Environmental Politics: Power and Principles in US Foreign Policy*, Ashgate, Aldershot, 2001.
39. See Robert L. Paarlberg, 'US Environmental Policy since Rio,' in Norman J. Vig and Regina S. Axelrod, *The Global Environment: Institutions, Law, and Policy*, Congressional Quarterly Press, Washington DC, 1999.
40. United Nations Conference on Environment and Development, *Report of The United Nations Conference on Environment and Development. Annex III. Non-Legally Binding Authoritative Statement of Principles for a Global Consensus on the Management, Conservation and Sustainable Development of All Types of Forests*, UNCED, Rio de Janeiro, 1992, pp. 5–7.
41. Nancy Alexander et al., *Marketing the Earth: The World Bank and Sustainable Development*, Friends of the Earth, Washington DC, 2002, available at www.foe.org.
42. Kenny Bruno and Joshua Karliner, *Earthsummit.biz*, Food First Books, Oakland CA, 2002.
43. Richard Elliot Benedick, *Ozone Diplomacy: New Directions in Safeguarding the Planet*, Harvard University Press, Cambridge MA, 1991, p. 121.
44. Marvin Soroos, *Endangered Atmosphere: Preserving a Global Commons*, University of South Carolina Press, Columbia, 1997, pp. 150–51.
45. *Alternative Fluorocarbons Environmental Acceptability Study*, at AFEAS www.afeas.org.
46. Ibid.
47. *Chemical Week*, 24 September 1997.
48. Fred Pearce, 'US Millers Fight for Banned Pesticide,' *New Scientist*, 5 October 2002, p. 11.
49. UNEP Technical Options Committee, *2002 Assessment Report of the Methyl Bromide Technical Options Committee*, United Nations Environment Programme, 2002, at www.teap.org; and WMO/UNEP, *Scientific Assessment of Ozone Depletion: 2002*, www.al.noaa.gov.
50. US Environmental Protection Agency, 'Questions & Answers About the MBr Critical Use Exemption,' www.epa.gov.
51. Sean D. Murphy, 'Prospective Liability Regimes for the Transboundary Movement of Hazardous Waste', *American Journal of International Law*, vol. 24, no. 31 (1994): 1561–70. Jennifer Clapp, 'The Toxic Waste Trade with Less-Industrialized Countries: Economic Linkages and Political Alliances,' *Third World Quarterly*, vol. 15, no. 3 (1994): 505–18; Sejal Choksi, 'The Basel Convention on the Control of Transboundary Movements of Hazardous Wastes and Their Disposal: 1999 Protocol on Liability and Compensation,' *Ecology Law Quarterly*, vol. 28, no. 2 (2001): 509–40.
52. Basel Action Network, 'The Basel Ban Amendments: The First Step Toward Environmentally Sound Management of Hazardous Wastes,' paper prepared for

the 16th Session of the Technical Working Group and 1st Session of the Legal Working Group of the Basel Convention, Geneva, 3–9 April 2000.

53. Gareth Porter, Janet Welsh Brown, and Pamela S. Chasek, *Global Environmental Politics*, Westview Press, Boulder CO, 2000.
54. Cited in Jim Puckett, 'The Basel Ban: A Triumph over Business as Usual,' Basel Action Network, 1 October 1997, available at www.ban.org.
55. Jim Puckett and Cathy Fogel, 'A Victory for Environmental Justice: The Basel Ban and How it Happened,' Greenpeace International, 1994, www.ban.org.
56. See Maria Isolda P. Guevera, 'The Basel Convention Export Ban Amendment: Arguments Against Ratification,' www.ban.org.
57. Puckett and Fogel, 'A Victory for Environmental Justice.'
58. Choksi, 'The Basel Convention.'
59. Ching-Ching Ni, 'Dante's Digital Junkyard: Chinese Laborers Eke Out a Living Using Acid, Fire and their Bare Hands to Recycle Mountains of Electronic Scrap, Most of it from the U.S,' *Los Angeles Times*, 6 April 2004.
60. Joyeeta Gupta, *Our Simmering Planet*, Zed Books, London, 2002.
61. Jeremy Leggett, *Carbon War: Global Warming and the End of the Oil Era*, Routledge, New York, 2001.
62. All quotations in this paragraph are from *Rocky Mountain News,* Denver CO, 12 December 1997, p. 68.
63. *Atlanta Journal and Constitution*, 7 May 1998.
64. Hugo Gurdon, 'The U.S. Should Unsign Kyoto,' *Wall Street Journal Europe*, 11 October 2002.
65. Union of Concerned Scientists, *Scientific Integrity in Policymaking: An Investigation into the Bush Administration's Misuse of Science*, UCS Report, February 2004.
66. *New York Times*, 5 June 2004.
67. *New York Times*, 19 June 2003.

chapter 7

1. Much of this section is inspired by the excellent discussion in Michael Maniates, 'Individualization: Plant a Tree, Buy a Bike, Save the World?' *Global Environmental Politics*, vol. 1, no. 3 (2001): 31–52.
2. Ibid.
3. William Baarschers, *Eco-Facts and Eco-Fictions: Understanding the Environmental Debate*, Routledge, London, 1996.
4. Maniates, 'Individualization.'
5. Marius de Geus, 'The End of Environmentalism? The Environment versus Individual Freedom and Convenience,' paper for the ECPR workshop, 'The End of Environmentalism?' Turin, 22–27 March 2002.
6. *All Things Considered*, National Public Radio, 23 June 2004.
7. Garrett Hardin, 'The Tragedy of the Commons,' *Science*, vol. 162 (1968): 1243–8.
8. Garrett Hardin, 'Extensions of "The Tragedy of the Commons",' *Science*, vol. 280 (1998): 682–4.
9. Political scientist Elinor Ostrom has uncovered a range of communities around the world that defy the tragedy scenario. Water producers in California voluntarily collaborated to prevent the destruction of the local water basin. Greek fishermen formed an ingenious cooperative effort to allocate fishing spots and so promote sustainable use of the local fisheries. Elinor Ostrom, *Governing the Commons: The Evolution of Institutions for Collective Action*, Cambridge University Press, Cambridge, 1990.

10. David T. Suzuki and Holly Dressel, *Good News for a Change: How Everyday People Are Helping the Planet*, Greystone Books, Vancouver, 2003.
11. Robert Gottlieb, *Environmentalism Unbound: Exploring New Pathways for Change*, MIT Press, Cambridge MA, 2001.
12. Suzuki and Dressel, *Good News for a Change.*
13. Robert D. Putnam, *Bowling Alone: The Collapse and Revival of American Community*, Simon & Schuster, New York, 2000.
14. Robet H. Dahl, *A Preface to Economic Democracy*, University of California Press, Berkeley, 1986.
15. Michael Daniel, 'Using the Fourteenth Amendment to Improve Environmental Justice,' *Human Rights: Journal of the Section of Individual Rights & Responsibilities*, vol. 30, no. 4 (2003): 15–18; Hilda I. Solis, 'Environmental Justice: An Unalienable Right for All,' *Human Rights: Journal of the Section of Individual Rights & Responsibilities*, vol. 30, no. 4 (2003): 5–7.
16. Luke W. Cole and Sheila R. Foster, *From the Ground Up: Environmental Racism and the Rise of the Environmental Justice Movement*, New York University Press, New York, 2001.
17. Jennifer Hattam, 'Why Race Matters in the Fight for a Healthy Planet,' *Sierra*, May–June 2004: 56–8.
18. Suzuki and Dressel, *Good News for a Change.*
19. See www.goldmanprize.org.
20. Norman Myers and Jennifer Kent, *Perverse Subsidies: How Tax Dollars Can Undercut the Environment and Economy*, Island Press, Washington DC, 2001, p. xvi.
21. Ibid.; Jane Holtz Kay, *Asphalt Nation: How the Automobile Took Over America and How We Can Take it Back*, University of California Press, Berkeley, 1997, pp. 120–26.
22. 'The Real Price of Gas,' report by the International Center for Technology Assessment (CTA), available at www.icta.org.
23. Lester Brown, *Eco-Economy: Building an Economy for the Earth*, W.W. Norton for the Earth Policy Institute, New York, 2001.
24. David Malin Roodman, *Paying the Piper: Subsidies, Politics, and the Environment*, Worldwatch Paper no. 133, December 1996.
25. *New York Times*, 12 May 2004.
26. Americans and Foreign Policy, Program on International Policy Attitudes (PIPA), available at ww.pipa.org.
27. Heikki Patomaki, *Democratising Globalisation: The Leverage of the Tobin Tax*, Zed Books, London, 2001.
28. Paul Hawken, 'Natural Capitalism,' *Mother Jones*, March–April 1997: 40–54.
29. Suzuki and Dressel, *Good News for a Change.*
30. John Feffer (ed.), *Living in Hope: People Challenging Globalization*, Zed Books, London, 2002.
31. Todd Gitlin, 'Shouts Bring Murmurs,' *Washington Post*, 16 April 2000.

Index